T0174073

BKSTS
Illustrated Dictionary of
Moving Image Technology

BKSTS
Illustrated Dictionary of
Moving Image Technology

Fourth edition

Martin Uren

Focal Press
Taylor & Francis Group

NEW YORK AND LONDON

First published 1983 as *BKSTS Dictionary of Audio-Visual Terms*
Second edition 1988
Third edition 1994
Fourth edition 2001

This edition published 2013
by Focal Press
70 Blanchard Road, Suite 402, Burlington, MA 01803

Simultaneously published in the UK
by Focal Press
2 Park Square, Milton Park, Abingdon, Oxon OX14 4RN

Focal Press is an imprint of the Taylor & Francis Group, an informa business

© Martin Uren 2001

All rights reserved. No part of this book may be reprinted or reproduced or utilised in any
form or by any electronic, mechanical, or other means, now known or hereafter invented,
including photocopying and recording, or in any information storage or retrieval system,
without permission in writing from the publishers.

Notices

Practitioners and researchers must always rely on their own experience and knowledge
in evaluating and using any information, methods, compounds, or experiments described
herein. In using such information or methods they should be mindful of their own safety
and the safety of others, including parties for whom they have a professional responsibility.

To the fullest extent of the law, neither the Publisher nor the authors, contributors, or
editors, assume any liability for any injury and/or damage to persons or property as a matter
of products liability, negligence or otherwise, or from any use or operation of any methods,
products, instructions, or ideas contained in the material herein.

British Library Cataloguing in Publication Data
Uren, Martin
  BKSTS illustrated dictionary of moving image technology .
  4th ed.
  1. Audio-visual equipment – Dictionaries   2. Broadcasting – Dictionaries
  I. Title   II. Dictionary of moving image technology   III. Moving
  image technology
  621.3′881′03

Library of Congress Cataloguing in Publication Data
BKSTS illustrated dictionary of moving image technology/Martin Uren –
4th ed.
  p. cm.
  Rev. ed. of: Dictionary of image technology . 3rd ed. c1994.
  ISBN 0-240-51632-X (alk. paper)
  1. Audio–visual equipment – Dictionaries.   I. Title: Illustrated dictionary of moving
  image technology.   II. Uren, Martin.   III. British Kinematograph Sound and Television
  Society.   IV. Title: Dictionary of image technology .
  TS2301.A7 B37 2000
  621.389′7′03–dc21   00–049036

ISBN 13: 978-0-240-51632-5 (pbk)

Composition by Genesis Typesetting, Laser Quay, Rochester, Kent

Contents

Preface to the fourth edition

It is 18 years since the first edition of this dictionary appeared as the *Dictionary of Audio-visual Terms* and this new fourth edition has become the *BKSTS Illustrated Dictionary of Moving Image T echnology.* The structure and content have changed to reflect the widening technological coverage and understanding needed by all those who work in or study the technical aspects of the moving image.

The content of this edition focuses on technologies relating directly to programme origination, production, distribution, transmission and display by whatever means, from the film set, computer graphics and multimedia to digital satellite broadcasting, cinema projection and DVDs. Within this context I have tried to expand the range of the dictionary by including more historical terms as well as a multitude of new ones. Ours is an industry that bristles with colloquialisms and I have tried to include a wide selection.

The book is divided into three main parts. Part 1 contains over 3300 definitions covering film, television, sound and multimedia technologies, together with technical terms from the computing, networks and tele- communications industries which are used in the modern converged media industry. Part 2 contains a quick look-up for nearly 700 acronyms. Almost all of these are more fully explained in Part 1. Part 3 is a series of 26 appendices which provide useful technical information across a range of topics.

In retaining the clear and concise style of earlier editions, together with an expanded range of appendices, I hope the *BKSTS Illustrated Dictionary of Moving Image Technology* is a useful supplement to the many technical publications covering our industry. With over 100 illustrations and frequent WWW references, the book is designed to appeal to a broad range of students and practitioners alike.

I should like to thank the BKSTS staff and my colleagues on the BKSTS Education and Training Committee and the Television Committee for their help and suggestions in preparing this new edition.

Martin Uren
Brighton, June 2000

Preface to the third edition

I was delighted to be asked to update the *Dictionary of Image Technology* which, since its first publication ten years ago, has become a standard reference book for the film and video industries.

Minimal change has been made to the style of the book, which remains as a memorial to Bernard Happé, and it is a tribute to him that the content has remained remarkably up-to-date in spite of the many changes which this industry has seen in the last five years.

I am grateful to John Croft, Richard Ellis, John Iles, Peter Owen and Arthur Piggott for their contributions and advice in the preparation of this edition.

Walt Denning
1994

Preface to the second edition

The past six years have seen continued development in all the technologies which we took as our original field and in preparing this fully revised edition we felt that its wide scope is better described as a *Dictionary of Image Technology* than by its previous title, *Dictionary of Audio-visual Terms*; in it, we have endeavoured to reflect both current practice and new proposals still being debated internationally.

As Editor I am grateful to all the members of the original committee who have contributed new material and to Stephen Lowe whose glossary in the BKSTS publication *Understanding Video* suggested several additions.

Bernard Happé, *Editor*
May 1988

Preface to the first edition

As its name indicates, the British Kinematograph Sound and Television Society is concerned with the practice of a wide range of technologies, and in undertaking the preparation of this dictionary of technical terms it was agreed that the work should reflect fully the extensive interests of the Society. We have therefore interpreted the term *audio-visual* in the most generous sense, embracing the preparation and presentation of pictures and sound by film and video as well as by tape-slide, film-strip and multivision.

Many of the boundaries which formerly divided photographic and electronic methods of recording and reproduction have tended to disappear during recent years and the interchange of these media now frequently provides an additional tool in the hands of the creative producer as well as extended facilities for exhibition. Motion picture film and still slides are incorporated in video productions, videotape is transferred to film, and tape-slide shows are converted to both film and videocassettes for alternative methods of distribution. There is thus ample justification for treating these varied disciplines within the same volume on a common basis.

In recognition of current developments, we have also considered it necessary to explain numerous terms in the field of computer practice, which now enters into many techniques of both production and presentation. The actual display of images in computer graphics and computer animation is an obvious example but computer -controlled operations in video and sound production, in animation photography and in motion picture processing are of increasing importance, while complex multi-screen shows involving dozens of projectors and hundreds of individual slides benefit greatly in both programming and presentation from the facilities available through microprocessors.

We hope therefore that by providing brief explanations of a number of such terms we shall help the understanding of these important applications. In these, as throughout the volume, we have endeavoured to introduce and define our entries on the basis of practical operations, since this is a dictionary of the usage of terms rather than a textbook or a technical encyclopedia. To the purist, some of these words and their applications may appear unusual or even questionable but it is the explanation of current employment rather than grammatical formality which has been our guide. We hope, in addition, that this publication will do something to assist in uniformity of interpretation, especially where the same term has acquired somewhat varied meanings in different contexts.

Our society is fortunate in being able to call upon so much expert knowledge in current procedures and practices and gratefully acknowledges the enthusiastic efforts of all the contributors, members of the Dictionary Editorial Committee, many of whom have previously prepared vocabularies as part of their own technical publicationsA wide range of other sources has been consulted to provide word lists and we are also indebed to John Halas, Brian Salt and many others, including industry associations, for the basis of many of the definitions which we have included.

In attempting to cover such a wide field of applied technology   , and moreover one which is still developing very rapidly   , omissions are inevitable and, despite our best endeavours, errors of definition or of usage may have passed uncorrected; readers ' comments and additions will be welcomed for inclusion in subsequent editions.

Bernard Happé, *Editor*

BKSTS A-V Dictionary Editorial Committee
Mike A. Ray (chairman)
Stanley W. Bowler
Hugh D. Ford
John Lewell
Jack Speller
Brian Watkinson
G. Ross Watson
David Wilkinson

April 1982

About BKSTS

Founded nearly 70 years ago, BKSTS   The Moving Image Society is the organization representing the interests of those who are creatively and technologically involved in the business of providing moving images in all areas of the media including film, television, cinema exhibition, multi-media and sound.

The society is dedicated to encouraging and promoting excellence in all aspects of film and television and achieves this by sustaining, educating and training both its members and non-members.

The society's members are drawn from all areas of the industry including senior management, operational, sales and marketing, cinematographers, editors, producers, multimedia, television and film.   There is roughly a 50–50 split between those operating in film to television.

The BKSTS has a flourishing network of regional branches and also welcomes overseas members from as far afield as Canada, Nigeria and Malaysia.

The BKSTS is supported by over 100 sponsor companies who provide both financial and practical support.Through the Diamond, Gold, Silver and Bronze sponsor packages, sponsor companies have the opportunity to promote themselves to the society's members and to the industry as a whole. They often contribute to the society's journal *Image Technology* as well as providing many of the members on council (the society's governing body).

A wide and diverse range of activities are provided for members including networking evenings, seminars, training courses, social events and conferences.

About the author

Martin Uren has over 20 years ' experience in broadcast television, beginning with BBC engineering, followed by nine years at ITN in engineering and training roles and, since 1993 as an independent broadcast training consultant. He delivers a wide range of technical and operational training courses, seminars and workshops for broadcast and media organizations and institutions. He has specialized in helping broadcast technicians and engineers keep up to date with the latest digital techniques and advances in broadcast related technologies.

His clients include the ITV companies, satellite broadcasters, and manufacturers. He has also run courses for the National Film and Television School and is a visiting lecturer on the degree courses at Ravensbourne College of Design and Communication, covering topics such as microwaves, satellite communications, fibre optics and digital transmission.

Martin was a member of the Skillset committee which developed S/NVQs for broadcast engineering. He is actively involved with the BKSTS, and is also a member of the R TS and the SMPTE.

For light relief Martin composes and arranges rock music and pursues many other interests including astronomy , sailing and genealogy . He is a member of Mensa and can also often be seen (and heard!) in the Brighton area on his Matchless vintage motorcycle.

Part 1

Definitions

A

A/B: **(1)** Originally monitoring audio signals after or before tape. Now used as a term for comparing audio signals reproduced by dif ferent systems. **(2)** A *cross fade* technique between two selected sources.

A & B cutting: A method of assembling original film material in two separate rolls, allowing optical ef fects to be made by double printing. See also *checkerboard cutting*.

A-B roll: **(1)** A linear video *editing* process where alternate shots are replayed from two *VTR* sources through a *vision mixer* and recorded onto a third *VTR*. This speeds up the *editing* process and allows *mixes* and *wipes* to be made between the two source shots. **(2)** A dual film projection technique where silent film is projected via one film chain while another uses *sound-on-film*.

A or B types: Used to identify the emulsion position in 16mm prints for projection. Type B is run emulsion to lens, type A emulsion to lamp.

A and B windings: The two forms of winding used for rolls of film perforated along one edge only . When unwinding clockwise the perforations are nearer the observer in A winding and away from the observer in B winding. Defined in standard SMPTE 75M. See **Appendix U** and **Figure A.1**.

aberration: Inherent deficiency of an optical system resulting in the formation of an imperfect image. The principal types are *astigmatism, barrel distortion, chromatic aberration, coma, curvature of field , pincushion distortion* and *spherical aberration*.

above ground level (AGL): Used in *microwave* link systems to describe the height of an *aerial* relative to the ground level at the foot of the tower.

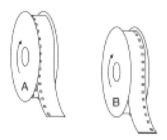

Figure A.1 A and B windings

absorption coefficient: The fraction of incident sound, light or *radio frequency* energy absorbed by a material at a given frequency and conditions.

absorption loss: In *telecommunications*, attenuation of the optical signal within an *optical fibre*, usually specified in dB/km.

Academy: The *Academy of Motion Pictur e Arts and Sciences* in the USA.

Academy aperture: The *aperture (2)* of a 35mm motion picture camera or *projector* with the dimensions specified by the *Academy*. Also called *Academy gate* or *Academy mask*.

Academy gate: See *Academy aperture*.

Academy leader: The *leader* on a film *release print* containing information and synchronizing marks as designed by the *Academy*.

Academy mask: See *Academy aperture*.

Academy of Broadcast Science (ABS): Part of the State Administration of Radio, Film and Television of China.

Academy of Motion Picture Arts and Sciences (AMPAS): See *Academy*.

acceptance angle: (1) Horizontal width of vision of a camera expressed in degrees. **(2)** The angular width of sound input to a *directional microphone* at which the sensitivity drops by 3 *decibels*.

access time: (1) The time taken to locate and read or write an item of information on a storage device such as a *hard disk drive*. **(2)** Time taken to find a *slide* in a *random access projector*, or a desired section on a *random access* audio or video reproducer.

access unit: *MPEG-2* term for the data block representing a data compressed audio signal (representing a few tens of milliseconds of digital audio) or video *frame (2)* (compare *presentation unit*).

accordion pleating: Film damage in the form of repeated folds.

accumulator: (1) A rechargeable *secondary cell* or *secondary battery*. **(2)** In a *microprocessor*, a *register (1)* for the storage and manipulation of instruction codes and data.

ace: A 1 kilowatt *Fresnel lens* spotlight.

acetate: (1) Cellulose triacetate, a clear transparent plastic used as the base for film and for *cels* in graphics. **(2)** A metal-cored *audio disc* on which the *modulation* is originally cut. (Hist.)

achromatic lens: A lens corrected for *chromatic aberration* during its manufacture.

Acquisition Numérique et Télévisualisation d'Images Organis ées en Page d'Ecriture: See *Antiope*.

action: (1) Command given by a film director for the performance of a scene to begin. **(2)** The performance of a scene in front of the camera. **(3)** The film recording the performance; the picture as distinct from the sound.

active component: An electronic component which uses a power supply to provide gain (compare *passive component*).

active format description (AFD): A *DTG* extension to the *DVB* digital transmission standard to indicate the 'area of interest' in the transmitted picture. Designed to help optimize the display on 4 :3 or *widescreen aspect ratio (1)* television *receivers*.

active matrix LCD: A *liquid crystal display* which uses an array of *thin film transistors* integrated into the display structure to switch the *pixels* making up the image.

active picture: The part of a video signal which carries the image to be displayed (compare *blanking period*). See **Figure I.2.**

active video: See *active picture*.

active region descriptor (ARD): In the UK's digital transmission system, this transmitted parameter is used by the digital *receiver* to enable the most appropriate display format for either 4 :3 or 16:9 images.

active window: In computers, the *window* in a *software application* that is currently selected for use.

actual sound: Sound recorded at the time of filming. Also called *actuality sound*.

actuality sound: See *actual sound*.

adapter: A means of interconnecting components of a system having different *terminations* or couplings.

adaptive decoder: One which optimizes its performance in response to changes in the input signal, such as an adaptive *comb filter* decoder for *composite video (2)* signals.

adaptive differential pulse code modulation (ADPCM): Form of digital coding and data compression using the dif ferences between successive *PCM* samples, not their absolute values. The system is often used with audio because of its low *latency (1)*. The amount of data compression is increased over *DPCM* by the *encoder* making a prediction of the next sample value based on previous sample(s). The difference between the prediction and the actual value is encoded. The *decoder* reverses this process to recover the signal.

adaptive transform acoustic coding (A TRAC): A form of digital audio data compression designed for *MiniDisc* by Sony under license from *Dolby*. It is also used in Sony' s *SDDS* multi-channel film sound format. It gives a data reduction of 5 :1 using *psychoacoustic* coding techniques.

additive colour: Colour mixed by a combination of light of the three *primary colours*: red, green and blue.

additive key: *Keying* in which two complementary video signals which have been shaped by multiplication with a *key signal* are added to create a combined image.

address: (1) A memory location in a *computer* containing an *instruction* or *data*. It may also identify a device connected to the computer . **(2)** A selected edit point during *time code editing*, usually identified by a *time code* number. **(3)** In data *networks*, the identification code for a *node* or device on the *network*.

address space: The available area defined by the co-ordinates of a *computer graphics* system.

advance: (1) In a motion picture print, the separation between the picture image and the corresponding point on the sound track required for correct *synchronization* in projection. See **Appendix G. (2)** The command for an automatic *slide projector* to advance to the next *slide*.

Advanced Broadcast Systems of Canada (ABSOC): A consultative group that has represented the interests of the Canadian television industry within the *Advanced Television Systems Committee (A TSC)* since 1990.

Advanced Communications Technology and Services (ACTS): The focus of the European Union's research effort to accelerate deployment of advanced communications infrastructures and services across the European Community.

Advanced Conversion Equipment (ACE): An early television *standards converter* developed by the *BBC* and manufactured by GEC. (T rade name.) (Hist.)

Advanced Digital Adaption Converter (ADAC): An early television *standards converter.* (Trade name.) (Hist.)

Advanced Intelligent Tape (drive) (AIT): (Sony). A data storage method based on 8mm *helical scanned* tape of 50 gigabyte capacity .

Advanced Research Projects Agency Network (ARP ANET): US Department of Defense network that eventually became the basis for the Internet.

Advanced Television (ATV): US term for high resolution digital television.

Advanced Television Systems Committee (ATSC): A committee formed in 1982 to establish voluntary technical standards for advanced television systems in the USA, including digital *high definition television.* Their 'Digital Television Standard' is based on *MPEG-2* video compression, AC-3 *(Audio Code 3)* audio and offers a wide range of video resolutions and audio services. See **Appendix S.** (Website: www.atsc.org).

Advanced Television Technology Center (ATTC): US non-profit making corporation that tests and recommends *hardware* solutions for delivery and reception of digital television and *high definition television* . (Website: www.attc.org).

Advertising Standards Authority (ASA): An authority set up in 1962 to ensure that non-broadcast advertisements appearing in the UK are legal, decent, honest and truthful. It is independent both of the advertising industry and government. (Website: www.asa.org.uk).

aerial: A device for receiving or transmitting *electromagnetic waves*, usually in the form of *radio frequency* signals. For transmission, an aerial converts electrical ener gy into electromagnetic ener gy, the opposite for reception. The American term in common use is *antenna*.

aerial image: An optical image formed in space rather than on a *screen (1)*. In aerial image photography, *titles* and other material may be combined with this spatial image.

AES/EBU interface: A digital audio standard defined jointly by the *AES* and the *EBU*. It defines an *interface (1)* for interconnecting professional digital audio equipment. The *interface (1)* carries 16-bit, 20-bit or 24-bit digital audio samples or non-audio data. Digital audio signals formatted to use this *interface (1)* are often simply referred to as AES/EBU audio.

A-format: A 25 mm (1 inch) *broadcast standard helical-scan reel-to-reel* analogue *videotape recorder* format that was never commercially successful but led to the development of *B-format* and *C-format* machines. (Hist.)

air: To transmit a television or radio programme.

air brush: A small compressed-air paint spray. The term is also used for a similar effect in *computer graphics*.

algorithm: A set of rules defining the steps used to perform a given task.

alias: (1) In analogue signal processing, interference between two closely related frequencies may produce unwanted dif ference frequencies not present in the original signals. These are aliases or *beat frequencies*. Visually the effect is called *moiré*. **(2)** In *digital signal processing*, an unwanted signal produced by a *sampling* process, usually associated with *analogue-to-digital conversion*. When the *sampling rate* is lower than twice the highest frequency to be sampled (i.e. below the *Nyquist frequency*) the reconstructed analogue output will contain dif ference frequencies not present in the original signal. These are aliases. See **Figure A.2**.

aliasing: The result of *sampling* at too low a rate for the *detail* in an image or high audio frequencies to be captured and reproduced correctly . In sound it can be heard as spurious low-frequency *noise*. In a video picture it appears as twinkling on sharp horizontal edges, or steps on diagonal line due to the *detail* of the image exceeding the *resolution* available on the *CCD imager* or the *raster* display device. See **Figure A.3**.

aliens: *Aliasing* effects, such as *ringing, contouring* and *jaggies* caused by interference between picture detail and a *raster* display. (Colloq.)

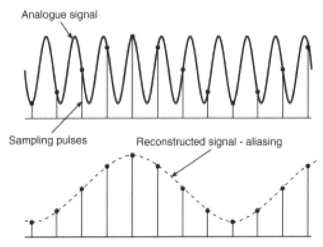

Figure A.2   Alias generation in analogue to digital conversion

alignment: The process of adjusting a system for optimum performance.
all pass: A *filter (3)* circuit which may alter   *phase* but not *frequency response* within a defined range.
alpha channel: See *key signal.*
alphanumeric: A set of characters which includes the letters of the alphabet and the digits 0 to 9.

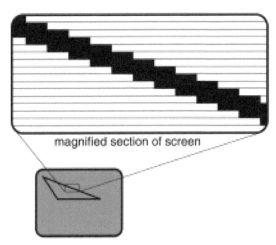

Figure A.3   Aliasing on a diagonal line

alternate: A facility in *slide* projection for rapid automatic changing between two projected images.

alternating current (AC): Electric current which reverses its polarity at regular intervals, e.g. the mains supply at 50Hz in Europe, 60Hz in USA.

Amalgamated Engineering and Electrical Union (AEEU): UK trade union formed by the merger of the *EEPTU* and the *AEU*.

Amalgamated Engineering Union (AEU): UK trade union that merged with the *EEPTU* to form the *AEEU*. (Hist.)

ambience: The acoustic combination of *reverberation* and *background (2)* in an environment.

ambient sound: The sound fed to the *loudspeakers* in the *auditorium* during presentations using multi-channel sound.

ambient temperature: The temperature of the surroundings of a device.

ambiophony: The creation of artificial *reverberation* in *auditoriums* by the use of time delays and multiple *loudspeaker* systems.

ambisonic: A system using multiple sound *channels (1)* to provide a three-dimensional effect. (Hist.)

American Broadcasting Company (ABC): US broadcast organization, which, after an earlier merger with United Paramount Theatres, became ABC, Inc. in 1996 when it merged with the Walt Disney Company. (Website: www.abc.go.com).

American Cinema Editors (ACE): An honorary society of film editors founded in 1950 to advance the craft of film editing. (Website: www.ace-filmeditors.org/home.htm).

American Institute of Electrical Engineers (AIEE): US body founded in New York in 1884 that merged with the *Institute of Radio Engineers* in 1963 to form the *IEEE*. (Hist.)

American National Standards Institute (ANSI): The principal standards development and approval body in the USA, founded in 1918. It represents the USA at the *ISO* and the *IEC*. (Website: www.ansi.org).

American Society of Cinematographers (ASC): US professional society for outstanding directors of photography founded in 1919.

American Standard Code for Information Interchange (ASCII): A standard code for transmitting data, consisting of 128 letters, numbers, symbols and special codes, each of which is represented by a 7-bit *binary* number. It was created in 1965 by Robert W. Berner.

American Standards Association (ASA): A rating of photographic film sensitivity, named after the association that developed it. It has been superseded by the *DIN (2)* rating. See also *exposure index* and *speed*.

amp: Abbreviation for *ampere*.

ampere (A): *SI unit* of electrical current, equal to one *coulomb* per second. Named after the nineteenth century French mathematician and physicist André-Marie Ampère. The term is often abbreviated to *amp*.

Ampex Digital Optics (ADO): Digital *special effects* device used in video *post-production.* (Trade name.) (Hist.)

amplifier: A device for increasing the strength of an input signal (in voltage and/or current) to provide greater power for subsequent use. The output signal remains in proportion to the input signal.

amplitude: The magnitude or intensity of a signal. In audio it is often expressed as *peak-to-peak* or *RMS* amplitude, in video as *peak amplitude.*

amplitude modulation (AM): Carrying information by varying the *amplitude* of a *carrier* signal on either side of its nominal value. See **Figure A.4**.

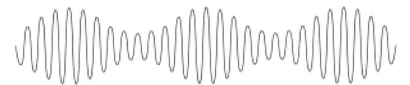

Figure A.4   Amplitude modulation

amplitude shift keying (ASK): Carrying digital information by varying the *amplitude* of a *carrier* signal between two fixed values.

anaglyph: A form of *stereoscopic* presentation in which the right and left eye images are reproduced in dif ferent colours, usually red and blue-green. The images are viewed using a *complementary colour* filter in front of each eye.

analogue component: *Component video* where the components are in their *analogue signal* form. Also called *component analogue video.*

analogue signal: A signal which varies continuously . Examples include the electrical output of a *microphone* or the voltage signal used to drive a *loudspeaker* (compare *digital signal*).

analogue-to-digital converter (ADC): A device which converts an *analogue signal* into its equivalent coded *digital signal*. Sometimes in video systems the conversion may involve decoding from an analogue *composite video (2)* source to *digital component* signals.

analysis projector: An apparatus for the detailed examination of a motion picture film record, frame by frame or at variable speed, often including dimensional measurement of the image.

anamorphic: (1) In a lens system, having different vertical and horizontal magnifications. **(2)** Used to describe how a *widescreen* video image is seen uncorrected on a 4 :3 picture display. **(3)** In *cinematography*, an

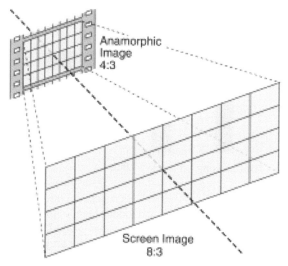

Figure A.5 Anamorphic image

image having lateral compression produced by an *anamorphic* lens. See *CinemaScope* and **Figure A.5**.

anechoic chamber: A room totally enclosed by non-reflecting surfaces. It may be designed for *audio* or *radio frequencies*.

Angstrom: Unit of *wavelength* measurement, equal to 10^{-10} metre. Named after the nineteenth century Swedish physicist Anders Jonas Ångström. (Hist.)

animate: (1) To produce the impression of movement by the rapid presentation of a series of still drawings or photographic images. (2) An instruction to begin the presentation of a dynamic visual event in programme production.

animatic: A limited *animation* using art work edited as a videotape storyboard and often used for test commercials.

animation: Producing the impression of movement in graphics or film *frames (1)* by recording and reproducing the images in rapid succession.

animation stand: A *camera mount* on a vertical framework or column, together with the *animation table* and its accessories.

animation table: A revolving table on an *animation stand*.

answer print: The first print of a completed motion picture production submitted for approval of the laboratory operations.

antenna: See *aerial*.

11

Figure A.6   Anti-aliasing on a curved line

anti-aliasing: Processing applied to a video signal to prevent or remove *aliasing*, particularly applied to *computer graphics* production. An example is the addition of *pixels* of intermediate grey levels and colour shades at the edges of text characters. See **Figure A.6.**

anti-alias filter: A *low-pass filter* used before an *analogue-to-digital converter* to prevent *aliasing*.

anti-halation: A coating or layer applied during the manufacture of photographic film to reduce *halation* in the exposed image.

anti-newton: *Slide mounts* with their glass surfaces treated to avoid the formation of *Newton's rings*.

antinode: A point at maximum motion in a vibrating system. See also *node (1)*.

Antiope: The French national *teletext* system *(Acquisition Numérique et Télévisualisation d'Images Organisées en Page d'Ecriture)*.

anti-phase: Out of *phase* by 180 degrees.

aperture: **(1)** In an optical system, the opening controlling the amount of light transmitted. See also *diaphragm (1)*. **(2)** In motion picture equipment, the opening at which film is exposed or projected. See also *gate*. **(3)** In a *rostrum camera* plate, the *exposure* opening. **(4)** In *slide* projection, the internal opening of a *slide mount*. **(5)** In a digital *filter (3)*, the number of *sampling* points used to define the *filter (3)* response. **(6)** In *satellite (2)* and *microwave* communication systems, the cross-sectional area of a parabolic *dish (1)* aerial.

aperture correction: An electronic system for increasing the apparent *resolution* of a video picture. Originally used to compensate for the finite *scanning* spot size in camera *tubes (2)*, now applied to *CCD imagers* and as an operational enhancement. If too much correction is applied, an unwanted black line appears around objects in the image and the *noise* in the picture increases. The process is also called *enhancement* and may be split into horizontal and vertical corrections.

aperture grille: A *cathode ray tube* for colour television displays using a mask of fine vertical slots to direct the *electron beams* to the appropriate coloured phosphor stripes, used in *Sony's Trinitron* range (compare *shadowmask tube*).

apogee: The point in the orbit of a *satellite (2)* when it is furthest away from the Earth.

apparatus room: A room near a television studio, *edit suite* or *computer graphics* area where the electronic equipment and video recorders associated with its operation are installed. Also sometimes called a *machine room*.

application: A *software* program which runs on a computer, but is not the *operating system* or a *programming* language or a utility.

application programming interface (API): **(1)** In computing, a set of *software* routines which gives programmers easy access to the services provided by the *operating system*. **(2)** In broadcasting, the *software* layer between the television *viewer (2)* and the interactive service provider for digital *broadcast* services.

applications specific integrated cir cuit (ASIC): An *integrated circuit* designed for particular rather than general applications.

apron: A forestage extending beyond the main stage into the *auditorium*.

Arbeitsgemeinschaft der Öffentlichrechtlichen Rundfunkanstalten Deutschlands (ARD): The first West German state *broadcast* television network, a consortium of nine regions.

architecture: **(1)** The internal structure of an electronic system, either *hardware* or *software*. **(2)** The physical interconnections and layout in a television studio.

archive: **(1)** A store with controlled environmental conditions for the long-term preservation of records, including film, audio tapes, *videotapes* and digital data. Also the contents of the store. **(2)** In computing, a collection of several *files* grouped together into a single *file* for transportation or long-term storage.

arena: An acting area surrounded by an audience.

array: **(1)** A series of *elements (2)* in an *aerial* system such as the *Yagi*. **(2)** A series of memory locations allocated a single name by a *computer program*.

arrayed waveguide grating (AWG): A *multiplexer (3)*/demultiplexer for using multiple *wavelengths* in an *optical fibre* data communication system.

Arri: **(1)** Arnold and Richter. German manufacturer of film cameras and lighting equipment. (Website: www.arri.net). **(2)** An Arriflex 16mm film camera made by Arri. (Colloq.)

artefact: (1) A visible defect in a video picture which is the result of a shortcoming in the processing the signal under goes, e.g. *cross colour* in *composite video (2)* . **(2)** An audible defect in a reproduced sound resulting from digital processing or compression of the sound.

artificial light: Light from a manufactured source, i.e. other than natural daylight, sunlight or moonlight.

artwork: Illustrative copy whether prepared by an artist, camera or other mechanical means. The term is also used in *printed circuit board* layout work.

ashcan: A 1000 watt *luminaire*. (USA, colloq.)

aspect ratio: (1) The ratio of picture width to height, often expressed with the height as unity, e.g. 1.85:1, or as 4:3 for television. A ratio of 16:9 has been adopted for *widescreen* television and *high definition television*. **(2)** The ratio of width to height of a *wipe (1)* pattern in a *vision mixer (1)*.

aspect ratio converter (ARC): A device for converting images for display in one *aspect ratio (1)* from a different source image format.

aspheric: A curved lens or mirror surface which does not conform to the shape of a sphere. It is used for correcting *aberrations* in optical camera lens and projection systems.

assemble editing: *Editing* where new video material, audio, *time code* and *control track* signals are joined onto the end of existing material, without introducing a visual disturbance at the edit point (compare *insert editing*).

assembler: A *computer program* which converts the low-level *mnemonic* instructions of *assembly language* to the *machine code* instructions required by a *central processor unit* to execute the *program*.

assembly language: A low-level *computer language* using *mnemonics* to assist in *program* development and understanding. Each *mnemonic* corresponds to a *machine code* instruction.

asset: Any material which a broadcaster or service provider can exploit, e.g. a programme, part of a programme or an image.

assistant stage manager (ASM): A member of a drama production team whose roles may include prompting, maintaining continuity and the wellbeing of the actors.

Association Française de Normalisation (AFNOR): The French national standards organization. (Website: www.afnor.fr/index_gb).

Association Internationale du Film d 'Animation (ASIFA): (The International Animated Film Society). The international association for makers of *animated* films and video productions. (Website: www.asifa-hollywood.org).

Association of British Theatre Technicians (ABTT): UK association formed in 1961, aiming to improve technical standards in the theatre by collating, discussing and re-distributing information on all aspects of performance and presentation. (Website: www.abtt.org.uk).

Association of Cinema and Video Laboratories Inc. (ACVL): US association founded in 1953 and concerned with improvements in technical practices and procedures and all other areas of interest to the laboratory industry. (Website: www.acvl.org).

Association of Imaging Technology and Sound (ITS): The largest US trade association serving the post production and digital media community. (Website: www.itsnet.org).

Association of Motion Pictur e Sound (AMPS): A UK or ganization of film sound technicians constituted to raise the standards and improve the professionalism of the craft of motion picture sound. (W ebsite: www.amps.net).

astigmatism: A lens *aberration* causing points away from the *optical axis* to be imaged as an oval spot or one of a pair of lines at different focal distances. It can be improved by *stopping down* the lens.

ASTRA: The *satellite (2)* system operated by *Société Européene des Satellites*. The first ASTRA *satellite (2)* was launched in 1988. ASTRA transmits over 500 analogue and digital television *channels (4)* and more than 380 analogue and digital radio *channels (4)*. (Website: www.ses-astra.com).

Asymmetric Digital Subscriber Loop (ADSL): Using twisted copper telephone lines to deliver a high data rate to users while a lower data rate return *channel (1)* is maintained.

asynchronous: (1) Sound which is not *synchronized* to the picture being presented. **(2)** Start-stop data transmission which avoids the need for a *synchronous* clock at the receiving end. **(3)** Two television signals of the same *scanning* standard but which are not *synchronized* to each other, i.e. not *genlocked.*

Asynchronous Transfer Mode (ATM): An *ITU* standard for the transfer of 53 byte fixed sized packets called *cells (3)*, suitable for voice, video and data and supporting *variable bit rate* transmission.

Atlantic Ocean Region (AOR): The area of the Earth centred on the Atlantic Ocean covered by a range of *satellites (2)* in *geostationary Earth orbit.*

ATM Forum: A global or ganization aimed at promoting *asynchronous transfer mode* within the broadcast and telecommunications industries. (Website: www.atmforum.org).

atmos: Abbreviation for *atmosphere*. See *background (2)*. (Colloq.)

atmosphere: See *background (2)*.

attack time: The time taken for a system (often audio) to respond to a change in input signal.

attenuator: A device to reduce the strength of an optical, acoustic or electrical signal (often audio or video).

attribute: A property or characteristic of a system. The term is often used in computing, but could apply, for example, to the *brightness* of a picture display device.

audio: Concerned with sound recording and reproduction, specifically a signal which carries sound information.

audio cassette: A *compact cassette* used for audio recording.

Audio Code 3 (AC-3): A *Dolby*® audio coding system for multi-channel sound. It was re-named *Dolby Digital*.

audio disc: A *gramophone* record. (Hist.)

audio distribution amplifier (ADA): An *amplifier* designed to give a substantially flat *frequency response* over the audio band and provide several separate output signals.

audio dub: Recording or re-recording the audio track(s) of a videotape recording without disturbing the existing video signal.

Audio Engineering Society (AES): A professional society devoted exclusively to audio technology. Its members include leading engineers, scientists and other authorities. It promotes standards in the professional audio industry (USA). (Website: www.aes.org).

audio follow video (AFV): An operational condition where an audio mixer and a vision mixer are interlocked so that when a video source is selected, the corresponding audio is simultaneously selected.

audio frequency (AF): Within the frequency range of human hearing, usually taken as 20 Hz to 20 kHz.

audio tape: Magnetic recording tape designed to record and reproduce *audio frequency* information, generally using an *analogue signal*.

Audio Video Interleave (A VI): A storage technique developed by Microsoft which combines audio and video into a single frame or *track (3)* on disk, saving space and keeping the audio and video *synchronized*. *Files* created in this way have an AVI *file extension*.

Audio-visual (A/V): Describes equipment, productions and presentations which combine sound and pictures.

auditorium: The part of a theatre, cinema or hall reserved for the audience.

Australian Broadcasting Corporation (ABC): The only national, non-commercial broadcaster in Australia. It was inaugurated in 1932 and has become Australia's largest broadcaster, entertainment and marketing organization. (Website: www.abc.net.au).

authoring: In *multimedia*, using a specialist *software* application *program* to combine text, *computer graphics*, *animation*, sound and video for games, presentations, interactive training etc.

auto-assemble: A video *editing* mode in which the *videotape recorder* rolls the tape back a short distance at the end of each recording and runs up to the start of the next to produce a perfect edit *transition*. See *backspace edit*.

auto-conforming: The final *on-line conforming* carried out under *computer* control and using an *edit decision list* which has been produced *off-line*.

Autocue: A type of television *prompter*. (Trade name.)

auto-cycling equipment: (1) In film *animation*, the automatic photography of one *cel* at specific intervals along the film, the intervening *frames (1, 2)* being skipped for subsequent *exposure*. **(2)** A video or audio tape recorder operating mode in which the tape is automatically rewound and replay starts again from the beginning or from a pre-set *counter* position.

auto-focus: (1) In a *projector*, an electro-optical device to ensure that the image is held in focus. **(2)** On an *animation stand*, the equipment that maintains focus as the camera field size is altered. **(3)** In a camera, a system that measures the distance to a *framed (6)* object and automatically adjusts the lens to focus the object.

automatic dialogue r ecording (ADR): See *automated dialogue replacement*.

automated dialogue replacement (ADR): Re-recording the dialogue for a film sequence in *post production*, often for reasons of clarity, missing dialogue or unwanted *background (2)*. Usually the actor watches the scene replayed on a screen and hears a series of beeps in a *headphone* giving a countdown to the beginning of the line. Also called *automatic dialogue recording*.

automatic double tracking (ADT): Artificial duplication of signals to simulate two or more musicians, etc.

automatic end stop: In magnetic recording equipment, detecting the end of a video or audio tape and shutting of f the equipment.

automatic frequency control (AFC): Adjusting a *receiver* to minimize unwanted change.

automatic gain control (AGC): Correcting input signal level variations to give an approximately constant output or the electronic circuit in an *amplifier* which carries out this process. It is often used to maintain video or audio levels in equipment where a hands of f operation is necessary. Also called *automatic level control*.

automatic lamp changer (ALC): For *slide projectors* with main and *standby* lamps.

automatic level control (ALC): See *automatic gain control*.

Automatic Noise Reduction System (ANRS): Trade name for an audio tape *noise reduction* system.

automatic repeat: Keyboard keys which automatically continue to repeat their action if held down.

automatic repeat request (ARQ): For automatic error control where the receiving device requests a copy of corrupted data to be re-transmitted. It is used where there is a return path from the receiving device to the transmitter, e.g. *modem* to *modem* links (compare *forward error correction (FEC)*).

automatic scan tracking (AST): Used in videotape replay to obtain perfect *tracking (4)* in slow-motion or *still frame* (Ampex). See *dynamic tracking*.

automatic switch-off: Switching off the power supply to a system by detecting an end-of-show signal, or by the end of the tape.

automatic track following (ATF): Signals used in rotary-*head (2)* video or audio tape reproduction to maintain accurate *tracking (4)*.

automatic voltage regulator (AVR): A system used to maintain a constant level of electrical mains supply to equipment.

automatic volume control (AVC): See *automatic gain control*.

auto-stop: See *automatic end stop*.

auto-threading: A film *projector*, tape recorder or similar device equipped to thread itself instead of having to be *laced* manually.

auto transition: In a *vision mixer*, a triggered transition of a specified duration where the motion of the *lever arm* is electronically simulated.

auxiliaries: Extra effects controlled during a *multivision* programme. The devices commonly controlled include film *projectors*, curtains, fountains, *lasers*, lights, motors, *pyrotechnics* and smoke machines.

available bit rate (ABR): An *ATM* traffic pattern where the data service is guaranteed a minimum data rate but may be allocated a higher rate if there is spare capacity on the network. See also *current bit rate (CBR)*, *uncommitted bit rate (UBR)* and *variable bit rate (VBR)*.

average picture level (APL): The average level of a video signal, often expressed as a percentage.

A-V format: A *compact cassette* track format with mono or stereo on tracks 1 and 2, and a *cue tone* or control on track 4. Track 3 is left unused to improve *cross talk*. This has been superseded by the *IEC* standardization of the use of tracks 3 and 4, and the space between them, for the *control track*.

Avid: Often used generically as the name of an electronic *non-linear editing* system, although actually a manufacturer of video *editing* and newsroom *computer*-based systems. (Trade name.)

A-weighting: Control over the *frequency response* of measuring equipment, commonly used for measuring ambient or electrical *noise*. The network is defined in *IEC* Recommendation 651.

azerty: A European alternative keyboard layout to *qwerty*, common in France and Germany. It has accents and features a comma rather than a full stop on the numeric keypad.

azimuth: (1) The angle between the gap in a magnetic *head (2)* and the direction of the tape motion. Similarly, the angle between the *slit (1)* in a photographic sound *head (3)* and the direction of film motion. (2) The compass heading of an *aerial* measured in degrees clockwise from north.

B

baby: A small spotlight.

baby legs: A short adjustable *tripod* used for low-angle shots. Called a *shorty* in the USA.

back coating: A thin conductive coating applied to the non-magnetic side of recording tapes to improve winding qualities, particularly at high speed.

back focus: The distance from the rear *element (3)* of a lens to the *image plane.*

background: (1) The main video picture source in a *keyer.* (2) Sound that defines the nature of a location. Called *atmosphere* or *atmos* in the UK. (3) The visible area in a *shot* behind the main action of the scene. (4) In *computer graphics* or *animation,* the image layer furthest from the viewer.

background light: Light shining on the *background (3)* of a shot. Also called *set light* or *dressing light.*

backing: (1) A coating applied to the base of photographic film to prevent or reduce *halation.* (2) The plastic base of magnetic recording tape.

backing-track: (1) A pre-recorded audio accompaniment. (2) Sound recording a combination of instruments and/or vocalists to support the main performer.

back light: Light directed from behind a subject towards the camera, to emphasize the subject's outline.

back porch: In an analogue video waveform, the period at *blanking level* between the end of the horizontal *sync pulse (2)* and the beginning of the *active picture* information. In *composite video (2)*, the *colour burst* is transmitted during this period. See **Appendix K.**

back projection (BP): The projection of a motion picture, still *slide* or video image on to the rear of a translucent *screen (1)* which is viewed from its front surface.

backspace edit: The join between shots on a video camera/recorder to maintain continuous video and control track information needed to edit the material together later. To maintain this continuity of information the tape winds backwards a short distance before the new recording begins.

backspacing: In videotape editing, rewinding a *videotape* from the desired in-point before an edit to give time for proper speed and synchronization to be achieved between the player and recorder .

backtiming: (1) In editing, calculating the in-point for an edit by subtracting the duration of the edit from the out-point. (2) US term for *genlock.*

baffle: A panel, usually of a *loudspeaker* cabinet, on which the individual *loudspeaker* units are mounted to improve radiation at low frequencies.

balance stripe: An extra stripe on striped magnetic film to ensure uniform winding.

balanced connection: A circuit consisting of a *ground* connection and two out-of-phase signal lines. These signal lines reject *interference* common to both connections (compare *unbalanced*).

balun: Abbreviation for balanced-unbalanced. An *impedance* matching and conversion component used when changing between a *balanced connection* and an *unbalanced* one. It is often used to convert between *twisted pair* and *coaxial cable* signalling connections.

banana plug: A single-pole plug, usually 4 mm in diameter.

band: A defined range of frequencies in the *electromagnetic spectrum*, e.g. *UHF* band or *Ku-band*. The *ITU* radio bands are listed in **Table B.1**. See also **Appendix N**.

banding: A *videotape recorder* defect in the *quadruplex* format which causes horizontal bands of different *hue*, *noise* or *saturation* to be visible on the output pictures. See also *head banding*. (Hist.)

band-pass filter (BPF): A *filter (3)* that passes a defined frequency range while rejecting higher and lower frequencies.

Table B.1   ITU radio band numbers

Band number	Initials	Band name	Frequency range
2	ELF	Extremely low frequency	30–300 Hz
3	ULF	Ultra low frequency	300–3000 Hz
4	VLF	Very low frequency	3–30 kHz
5	LF	Low frequency	30–300 kHz
6	MF	Medium frequency	300–3000 kHz
7	HF	High frequency	3–30 MHz
8	VHF	Very high frequency	30–300 MHz
9	UHF	Ultra-high frequency	300–3000 MHz
10	SHF	Super high frequency	3–30 GHz
11	EHF	Extremely high frequency	30–300 GHz
12	THF	Tremendously high frequency	300–3000 GHz

band-stop filter: A *filter (3)* that rejects a defined range of frequencies while passing higher or lower frequencies.

bandwidth: The range of frequencies passed by a device. It is usually specified as the range between the limits at which the response falls by 3 dB from the maximum.

bandwidth-segmented transmission (BST): A derivative of *OFDM* used in Japan for their *ISDB* digital television services.

bar: See *barrel*. (Colloq.)

barcode: A number of characters encoded as thin black and white stripes and printed on labels, e.g. on *videotape cassettes* in a library system. See also *keycode*.

barker channel: A *satellite (2)* or *cable television channel (4)* used to promote programming on other *channels (4)*.

barndoors: Hinged flaps which restrict the light beam from a *luminaire*.

barndoors wipe: A *wipe* in the form of opening or closing doors. See **Figure W.1**.

barney: A soft padded cover used to reduce the noise of a motion picture camera while still allowing hand holding and portability .

barrel: A metal tube, usually 48mm diameter , from which *luminaires* or scenery can be suspended. Also called a *bar*.

barrel distortion: Image distortion in an optical or video system causing a rectangle to appear to have convex sides and compressed corners (compare *pincushion distortion*). See **Figure B.1**.

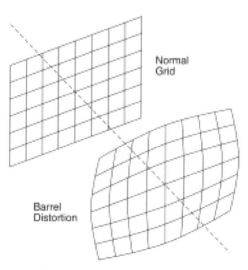

Figure B.1 Barrel distortion

barrier strip: A type of electrical connector comprising a series of screw terminals mounted on, or in, an insulating strip.

bars: See *colour bars.*

base: For film or tape, the flexible support on which a photographic emulsion or magnetic coating is carried.

baseband: In transmission, a signal in its original form before *modulation* or after *demodulation.*

basher: A small studio *luminaire* placed close to the camera or attached to it. Also called a *camera light.*

batch number: Identification of a quantity manufactured at one time with uniform characteristics, particularly film *raw stock.*

bath: A processing solution for photographic film, especially a *developer,* or its container tank(s).

battery back-up: A battery system used to prevent any loss of data when power to a *computer,* or other system, fails or is turned of f.

baud rate: In digital transmission, the number of information carrying *symbols* transmitted each second. Except for a few simple *modulation* formats, the baud rate is less than the *bit rate.* Named after the nineteenth century French engineer Jean Maurice-Emile Baudot.

bayonet mount: A push-and-twist type of *lens mount* for cameras.

Bayonet Neill-Concelman (BNC): Type of locking connector for *coaxial cable,* also known as a Bayonet Normalized Connector , British Naval Connector, Bayonet Nut Coupler or Baby N Connector .

bazooka: An adjustable *monopod* to take a *pan-and-tilt head (4).*

B-channel: A 64 kbit/s digital bearer *channel (1)* in *Integrated Services Digital Network* systems.

beaded screen: A directionally reflective projection *screen (1),* whose surface is composed of minute glass or plastic spherical beads.

beam: The signal transmitted from a *satellite (2)* or an *uplink* from an *earth station.*

beamwidth: The angular width in degrees of the conical *beam* radiated by a *dish (1) aerial.*

bearding: A visual defect in *videotape* reproduction which appears as a series of short random horizontal black lines to the right of high contrast dark to light edges. It is often caused by worn tape *heads (2)* or improper adjustment of a *videotape recorder.*

beat frequency: A spurious frequency produced by the combination of two signals at slightly dif ferent frequencies. See *alias* and **Figure A.2**.

behind the lens (BTL): A *filter (1)* position in a camera.

bel: Unit used to compare electrical power levels, sound intensities, etc. For example, the power dif ference between two signals W_1 and W_2 is $\log_{10} (W_1/W_2)$ bels. One bel is therefore a power increase of 10 times. The more commonly used form is the *decibel* (dB, one tenth of a bel), so 10 dB is a factor of 10 and 3 dB is a factor of 2. Named after the American inventor of the telephone, Alexander Graham Bell.

best boy (or girl): In film production, the assistant chief lighting technician or second electrician, the first assistant to the *gaffer*.

Betacam: A *videotape* recording system which uses 12.5mm ($\frac{1}{2}$ inch) tape and analogue component recording of *luminance* and *colour difference signals* on separate tracks. (Sony trade name.) (Hist.)

Betacam SP: *Betacam* Superior Performance. A *broadcast standard high band* development of *Betacam*.

Betacam SX: A *digital videotape recorder* format from Sony that uses 12.5mm ($\frac{1}{2}$ inch) tape and *MPEG-2* video data compression. The format uses a two-*frame (2) GOP* using alternate *I-frames* and *B-frames* which makes it possible to *edit* at any *frame (2)* boundary.

Beta-format: The tape format used in the *Betamax* system. (Hist.)

Betamax: A consumer quality *video cassette recorder* system using 12.5mm ($\frac{1}{2}$ inch) tape. Launched by Sony in 1975. (Trade name.) (Hist.)

Betamovie: A *camcorder* using the *Betamax* tape format. (Hist.)

B-format: A 25mm (1 inch) *broadcast standard helical-scan reel-to-reel* analogue *videotape recorder* system. (Hist.)

B-frame: Bi-directionally coded frame. An *MPEG* compressed picture in a video sequence which can take into account redundancy between itself and the previous or next *I-frame* or *P-frame* in the *GOP* sequence. *Macroblocks* of the picture to be coded are compared with the corresponding ones (which may be shifted using *motion estimation*) in the previous or next picture to find the one with least changes (i.e. most *redundancy*). This makes the B-frame the most efficient at coding images.

biamp: Two *amplifiers* used to drive the low and high frequency drivers of a two *loudspeaker* system. The *amplifiers* are preceded by an electronic *crossover network*.

bias: In audio tape recording, an *ultrasonic* signal added to the signal to be recorded. The combination is applied to the record *head (2)* to reduce distortion.

bias trap: A *high frequency (2) filter (3)* to remove the unwanted high frequency *bias* from the output or metering circuits of an *audio tape* recorder.

bi-directional microphone: One with nominally equal sensitivity to sound arriving from front or rear. It will have a figure-of-eight *polar diagram*. See also *figure-of-eight microphone* and **Figure D.5.**

big close-up (BCU): Generally showing the head of a subject.

bilateral sound track: A variable area photographic sound track whose *modulation* is symmetrical about its centre line. It may be single, double or multiple.

Bildschirmzeitung: A German *teletext* system.

bimorph: A sandwich of two barium titanate crystals used in the construction of *dynamic tracking heads* in *videotape* replay. An applied voltage causes one half to shrink thus causing the bimorph to bend. Two bimorphs are typically used to maintain accurate *zenith* by a parallelogram action. See **Figure B.2** and **Figure D.6**.

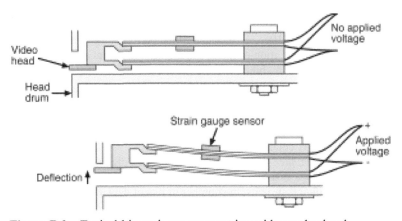

Figure B.2   Typical bimorph arrangement in a video replay head

binary: A system of numbering which uses only two states. These are usually expressed as 1 or 0 (one or zero).

binary phase shift keying (BPSK): *Phase modulation* with two states used in some data communications systems.

binaural sound: Sound recorded for *headphone* listening using two *channels (1)* which have been generated by modelling the human hearing process. Two *microphones* are placed in the same relative positions as human eardrums, using a 'dummy head'. This gives the listener the closest impression of being in the position occupied by the dummy head.

binder: The carrier for the magnetic particles in the coating of a magnetic recording tape.

binky: In motion picture sound, a mixing *premix* sheet showing the layout and content of *premixes*, often with one column for each *premix*. (Colloq.)

binocular vision: The use of both eyes in viewing a scene to extract *stereoscopic* information about the scene.

bipack: Two films run in contact through a motion picture or *rostrum camera*. Both may be *raw stock* simultaneously exposed or one may be exposed through a processed image on the other .

biphase: An electronic signal which gives film speed and direction information.

biphase mark: A method of *channel coding* which removes any *DC* component, is *self clocking* and has a limited *bandwidth*. It is used to encode *longitudinal time code*. It is not the same as *Manchester code*.

bird: A *telecommunications satellite (2)* located in Earth orbit. (Colloq.)

bit: A single Binary digIT. It can only have a value of 0 or 1, usually represented by two different voltage levels.

bit at unit density: Alternative derivation of *baud*.

bit error rate (BER): A measure of the performance of a digital system. The proportion of the bits in a data-stream that are corrupted in transmission. A bit error rate of 10^{-6} means that there is an average of one error per million bits transmitted (compare *errored second*).

bit error rate test (BERT): A data communications test in which a known data pattern is inserted at a *transmitter* and the number of errors received at a *receiver* is counted.

bit plane: A *computer graphics* store for holding a digital representation of a display image as a pattern of bits.

bit rate: In digital systems, the number of bits of data transmitted each second.

bit rate reduction (BRR): A combination of mathematical processing techniques used to decrease the data transmission and storage requirements of digital video, audio or data signals.

black and burst: See *colour black*.

black and syncs: See *colour black*.

black-and-white (B/W): Refers to *monochrome* materials and film or video images.

black body: A theoretical perfect radiator of energy whose colour when heated corresponds to the *colour temperature* of a matching source.

black burst: See *colour black*.

black crushed: A picture condition where a reduction in *black level* results in loss of gradation in dark shadow areas.

blacking: Recording a *control track*, *time code*, black video and silence onto a blank *videotape* to prepare it for *insert editing* operations. Also called *striping (3)*.

black level: The *amplitude* of a video signal representing zero *brightness*, i.e. the darkest part of the picture. See **Appendix K**.

black level clamp: An electronic circuit which holds the *black level* of a video signal at a fixed voltage level.

black slice: See *noise slice.*

blade: A narrow *flag (1)* to control part of the light beam from a *luminaire.*

blanked vector: In a vector-based line-drawing display, a vector having no intensity, used to change the current position of the beam without drawing a line. See *vector graphics.*

blanking level: The nominal video signal level during the *blanking period.* See **Appendix K.**

blanking period: The black frame around the television picture between the *active picture* and the overall video signal. (See **Figure I.2.**) Under normal circumstances this area is not visible since the displayed picture is enlarged to slightly overfill the display device. The purpose of blanking is to isolate the picture from the *scanning* and *synchronization* processes. The blanking period also contains the negative-going *sync pulses* (see **Appendix K**). Part of the blanking period above the picture may be used for *Teletext*, test signals or *VITC*. See also *horizontal blanking period, field blanking* and *vertical interval.*

blasting: An explosive distortion caused by the wind effect of breath on a *microphone.*

bleep: A short audible signal of medium pitch.

bleeper: An electrical or mechanical means of producing a *bleep* signal.

blender: A facility in *Quantel's* most sophisticated digital *computer graphics* equipment for video *mixing, keying* and *colour correction.*

blimp: A solid soundproof housing for a film camera.

blind wipe: A *wipe* in the form of opening venetian blinds. See **Figure W.1.**

blinking: Alternately displaying and not displaying a *title* or object on the *screen (1)*, usually to draw attention to it.

block: (1) A group of characters, words or records treated as a single unit in a *computer* storage system or in a transmission system. (2) A group of *pixels*, usually 8 horizontal by 8 vertical, in a digital picture used as a basic unit within a *BRR* system. (3) The minimum number of bytes that can be written to or read from a disk drive. See **Figure H.2.**

blockiness: A picture distortion in a video data *compression (2)* system where the edges of *blocks (2)* become visible, either as the result of a fault condition or where input picture complexity overwhelms the *compression (2)* system. Also called *blocking.*

blocking: (1) Saturation of an electronic circuit which prevents the passage of signals. (2) See *blockiness.* (3) The tendency for adjacent layers of magnetic tape on a *spool (1)* to stick together.

Blond: A variable-beam floodlight, typically 1 kW or 2 kW. (Trade name.)

blooming: (1) Coating the surfaces of a lens to reduce reflection losses. **(2)** The defocusing of areas of a video picture that are over -bright.

blooping: Applying a special opaque ink, paint or tape. On a photographic sound-track it is a triangular opaque patch to eliminate the noise caused by a join.

blow up: To enlarge photographically.

blue backing shot: In special effects, foreground action shot against a uniform blue *background (3)* for combination with another scene by *travelling matte* or *chromakey* processes.

Blue Book: (1) The specification for the *DV* common digital interface (*IEC* 61834). **(2)** The specification defining the Enhanced Music CD (*CD Extra*) for audio and data.

B-multiplexed analogue component (B-MAC): An analogue component television transmission system. See *multiplexed analogue component*. (Hist.)

BNC: Type of bayonet locking connector for *coaxial cable*. See *Bayonet Neill-Concelman*.

board: (1) An assembly of electronic components on a single support. **(2)** A lighting control panel.

boilerplate: US term for *cut and paste*.

bonding: In *ISDN*, joining together two *B-channels* to achieve a 128 kbit/s data transfer rate. Also called aggregation.

boom: A long *microphone*, camera or lighting support used nominally parallel with the floor . It may be telescopic and is usually held by a vertical stand.

boot: (1) To put a *computer* system into an operating condition. **(2)** A protective cover for electrical connectors.

border: A coloured or *monochrome* edge to a *key* or *wipe pattern*.

boresight: The line in the direction of maximum gain in an *aerial* system, particularly *satellite (2) dish (2) aerials*.

bounce: (1) See *hop*. **(2)** In television, a short-term periodic variation of *black level*. **(3)** See *contact bounce*. **(4)** To direct light to reflect from a ceiling or other large surface, usually white.

bouquet: A suite of digital *services* in a *DVB* transmission system. The grouping can be used by the *set-top box* in the way it presents the various *services* to the user . A *service* can be present in one or more bouquets.

bouquet association table (BA T): Optional *MPEG-2* data table which identifies the *services* in a *bouquet*. See **Appendix O**.

box: The volume of space in which a *satellite (2)* must be maintained by *stationkeeping* operations while in a *geostationary Earth orbit*.

box wipe: A *wipe* in the form of a square or rectangle. See **Figure W.1**.

braiding: A collection of fine wires which are plaited together to form a flat cable or to act as a *screen (3)* in *coaxial cable* or *multicore (1)* cable.

branchaloris: A tree branch held in front of a *luminaire* to create a foliage effect. See also *cookie (1)*.

breakdown: In motion picture laboratory practice, the separation of usable *takes (1)* from the rolls of processed original camera negative.

breakthrough: Unwanted signals appearing in a circuit from other circuits or devices.

breakup: A severe disturbance on an audio or video signal, perhaps due to loss of *sync* or tape damage.

breathing: **(1)** In sound recording, a distracting effect in a *compressor (1)* where the *background (2) noise* level varies with the level of the desired signal. It is also called *pumping.* **(2)** A defect in a motion picture camera, printer or *projector* where repeated movement of the film in and out of the correct plane causes variations in image *focus.*

breezeway: US term for the period in a video waveform between the *trailing edge* of a *sync pulse* and the start of the *colour burst.* See **Appendix K.**

bridge: **(1)** An electrical or electronic instrument for measuring resistance, capacitance or inductance. **(2)** A device linking two similar data *networks.* This allows *traffic* on the networks to be segregated, which improves the overall efficiency of data transfers across the system.

bridging: Electrically joining two or more points.

brightness: **(1)** The property of a surface emitting or reflecting light. In lighting, it is our impression of the amount of light received from a surface. It is measured in *candelas* per square metre and is called *luminosity* in the USA. **(2)** In video signals, the instantaneous level of *active video* conveying the corresponding brightness information.

brightness control: In television picture display equipment, the control which adjusts the *black level.* It is adjusted to make the black areas in the picture appear black when viewed in different room lighting conditions.

British Academy of Film and Television Arts (BAFTA): The UK's leading organization promoting and rewarding the best in film, television and interactive media. It began in 1947 as the British Film Academy, whose membership were the most eminent names in British film, and became BAFTA in 1957. (Website: www.bafta.org).

British Approvals Board for Telecommunications (BABT): Part of the *British Electrotechnical Approvals Board.*

British Board of Film Classification (BBFC): Set up by the film industry in 1912 as the British Board of Film Censors, the BBFC is an

Table B.2 UK film classifications

U	UNIVERSAL	Suitable for all.
PG	PARENTAL GUIDANCE	Some scenes may be unsuitable for young children.
12		Suitable only for persons of twelve years and over (introduced on video 1 July 1994).
15		Suitable only for persons of fifteen years and over.
18		Suitable only for persons of eighteen years and over.
R18	RESTRICTED DISTRIBUTION	Only available through specially licensed cinemas or sex shops to which no one under eighteen is admitted.
Uc	UNIVERSAL	Particularly suitable for young children (an additional category asked for by the video industry to be used for works stocked on the children's shelves of video shops).

independent UK non-governmental body , deriving its income from the fees it char ges for its services. Its responsibilities include the classification of films and *videos (3)*. See **Table B.2**. (Website: www.bbfc.co.uk).

British Broadcasting Corporation (BBC): UK public service broadcast organization formed in 1927 from the British Broadcasting Company which had been established in 1922. (W ebsite: www.bbc.co.uk).

British Electrotechnical and Allied Manufacturers' Association (BEAMA): The largest trade association federation in the UK, providing technical, legal, overseas marketing, standards, economics and commercial services to all the major manufacturing companies in the electrical and allied industries. It acts as an interface between industry, government departments, European Commission, etc. (W eb-site: www.beama.org.uk).

British Electrotechnical Approvals Board (BEAB): An independent UK national approvals body formed in 1960. It operates a safety testing and approval scheme for electrical and electronic equipment, measuring instruments, domestic appliances, and for controls used in household equipment in accordance with recognized national, European and/or international safety standards. Approved products can carry the BEAB APPROVED mark. (Website: www.beab.co.uk).

British Federation of Audio (BFA): UK organization formed in 1965 by hi-fi equipment manufacturers as the Federation of British Audio, becoming the British Federation of Audio in 1994. Its members are now manufacturers and distributors of a wide range of audio, home-theatre products and accessories. (Website: www.british-audio.org.uk).

British Film Commission (BFC): Founded in 1991 to promote the UK as a production centre for world movie making. (W ebsite: www.bfc.co.uk).

British Film Institute (BFI): UK national agency set up in 1933 with responsibility for encouraging the arts of film and television and conserving them in the national interest. (W ebsite: www.bfi.org.uk).

British Industrial and Scientific Film Association (BISFA): See *International Visual Communications Association.* (Hist.)

British Interactive Media Association (BIMA): UK trade body for the multimedia industry. (Website: www.bima.co.uk).

British Kinematograph, Sound and Television Society (BKSTS): A professional society whose interests include the technical aspects of film, sound, television and audio visual in all branches of entertainment, education, science and industry . Its preferred title is BKSTS – The Moving Image Society. (Website: www.bksts.com).

British Radio and Electr onic Equipment Manufactur ers' Association (BREMA): UK trade association aiming to promote the consumer electronic manufacturing industry in the UK, ensuring that a high standard of quality , design and workmanship is maintained in the industry. (Website: www.brema.org.uk).

British Standard (BS): The prefix for standards generated by the *BSI.*

British Standards Institution (BSI): The British national standards authority. It represents the UK in the *IEC, ISO* and *CENELEC.* It has a technical committee which co-ordinates the views of UK members to *ETSI* and therefore works closely with the *DTI.* (Website: www.bsi.org).

British Summer Time (BST): The standard time adopted in Britain between March and October , one hour ahead of *Greenwich Mean Time.*

British Telecom (BT): The largest UK national *telecommunications* services provider. BT Broadcast Services supplies a range of terrestrial, *satellite (2)* and *satellite (2)*-based multimedia services for international television and radio broadcasters. (W ebsite: www.broadcast.bt.com (Broadcast Services)).

British Universities Film and Video Council (BUFVC): UK representative body founded in 1948 as the British Universities Film Council aiming to promote the production, study and use of film and related media in British higher education and research. (W ebsite: www.bufvc.ac.uk).

British Video Association (BVA): UK trade body founded in 1980 representing the interests of publishers and rights owners of video home entertainment. It liaises with government, the media, other industry bodies and carries out extensive market research. (W ebsite: www.bva.org.uk).

broad: A *luminaire* for wide general illumination.

broadcast: The transmission of television, radio or other services simultaneously to a large group of *receivers*.

Broadcasters' Audience Research Board Limited (BARB): The main source of television audience research in the UK, created in 1980 when the *BBC* and ITV decided to have a mutually agreed source of television audience research. BARB became operational in 1981. (W ebsite: www.barb.co.uk).

Broadcasting, Entertainment, Cinematograph and Theatre Union (BECTU): Independent trade union for those working in broadcasting, film, theatre, entertainment, leisure and allied industries who are primarily based in the UK. It represents permanently employed, contract and freelance workers within these industries. (W ebsite: www.bectu.org.uk).

Broadcasting Standards Commission (BSC): UK statutory body for standards and fairness in broadcasting. It covers all television and radio, including BBC and commercial broadcasters as well as text, cable, *satellite (2)* and digital services. (Website: www.bsc.org).

broadcast standard: In video practice, representing the quality of recording and reproduction which meets the stringent requirements of *broadcast* organizations and regulating authorities.

Broadcast Video U-matic (BVU): A videotape recording system using 19mm ($\frac{3}{4}$ inch) tape. (Sony Trade name.)

broadcast wave file (BWF): *File* format defined by the EBU which contains the minimum information considered necessary for all broadcast applications.

broad pulses: The group of five *pulses* which make up the vertical sync pulse in the *vertical interval*. See **Appendix L**.

brouter: A data *network* device which acts as a *router* if it recognizes the *protocol* of the transmitted data or a *bridge* if it does not.

brown out: A US originated term for a mains voltage reduction which is liable to lead to equipment malfunction.

browse station: A video *workstation* on a *network* which can be used to look at stored images or video clips. Browsing implies examining but not altering.

Bruch blanking: The sequence of switching off the *colour burst* on certain lines during the *vertical interval* in a *PAL* system to improve colour stability in a colour television *receiver*. It ensures that each field starts with the same *colour burst* phase as that at the end of the previous field. It is named after a German, Walter Bruch.

brush: **(1)** A wiping electrical contact commonly found in electric motors and generators which comprises a carbon compound conductor which is pressed onto a smooth conducting surface by a spring. **(2)** In *computer graphics*, a marker that draws a line or a pattern.

brute: A large high-intensity spotlight, usually an arc lamp.

bubble: Any form of *luminaire.* (Colloq.)

buckle switch: A switch in a motion picture or *rostrum camera* to stop it running if the film piles up or is improperly threaded.

buffer: **(1)** A digital storage device which receives data and then later releases it. It can be used to convert a *variable bit rate* input to a *constant bit rate* output. **(2)** An analogue electronic device which isolates one part of a *circuit* from another. **(3)** A protective covering for an *optical fibre* cable.

bug: A *software* error which causes a system to behave erratically or stop altogether. The term originates from the days when computers used thermionic valves and relays. Insects were attracted into the equipment by the heat and light and caused problems. See also *debugging.*

build up: **(1)** In film *editing*, blank spacing inserted to represent missing sections. **(2)** A term used for illuminating a row of lights or images, where a new one is added to one end of the row without removing any of those already there.

bulk eraser: Equipment which uses powerful magnetic fields to remove previously recorded signals from *magnetic tape* or film when wound in a complete roll, as in a cassette.

bulk storage: A storage medium designed for lar ge quantities of data.

bulletin board service (BBS): An electronic message-passing service.

burn: **(1)** A defective area in a video camera *tube (2)* caused by focusing an excessively bright light on it. (Hist.) **(2)** A permanent image left on the phosphor surface of a *cathode ray tube* often due to displaying a constant image over a long period of time.

burn-in: **(1)** A technique for producing white or light coloured lettering, logos, etc., by over-*exposure* on to an exposed image in the camera (Hist.) or by superimposed projection. **(2)** The addition of *time code* numbers visible in the picture of a *videotape* record. **(3)** A pre-conditioning technique for components or equipment to improve reliability

burnt in time code (BITC): The *time code* numbers keyed into a dubbed *videotape* for viewing and logging purposes.

burst: See *colour burst.*

burst flag: A *pulse* in a *composite video (3) encoder* which switches unmodulated *subcarrier* onto the output at the appropriate time to form the *colour burst.* Also called *burst gate.*

burst gate: See *burst flag*. Also a *pulse* in a *composite video (3) decoder* derived from the incoming *sync pulses* which allows the *colour burst* to lock up the *decoder's subcarrier* generator.

bus: (1) A row of switch buttons on a vision mixer that allow selection from the available video sources. **(2)** An internal electrical connection between different parts of a system. Buses can be serial (e.g. *USB*) or parallel (e.g. *parallel port*). They can carry power , data, address and control information.

bus network: A *network topology* where all the devices share a single cable. *Ethernet* can use this *topology.*

butterflies: Very large *scrims (1)* used on location lighting.

butt splice: A join in a film or tape in which the two ends are not overlapped. When the ends are fused together , it is termed a butt weld.

buzz track: (1) A test film with a special photographic sound track used to determine the correct central position of the *scanning* beam *slit (1)* of an optical sound reproducer. Different frequencies are used on its opposite edges to check correct centring in the reproducer. See **Figure B.3**. **(2)** A sound track recording of local *background (2)* only, used to fill gaps in commentary or dialogue.

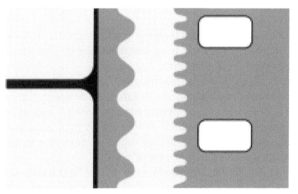

Figure B.3   Buzz track

B-weighting: A *frequency response* weighting network used for measuring *noise*. The network is defined in *IEC* Recommendation 651 'Precision sound level meters'.

byte: A group of eight *bits*.

C

Cable Communications Association (CCA): Trade body for the UK cable communications industry.

Cable Television Association (CTA): UK organization representing the views of the *cable television* companies.

Cable Television (CATV): A system for distributing television signals by means of underground or overhead cable links to individual subscribers' homes or small businesses. See also *hybrid fibre-coax.*

cache: (1) In computers, a fast memory used as an intermediate store between the main memory of a *computer* and the *central processor unit.* It is used to improve the overall processing speed of the *computer.* **(2)** In digital video systems, a *digital disk recorder* or *RAID* system storing programmes, commercials, promotions, etc. before transmission.

calender: To polish a coated surface, for example, *magnetic tape.*

calibration tape: A pre-recorded *magnetic tape* containing specific test signals at standardized levels.

camcorder: A compact hand-held video camera with an integral *videotape recorder.*

cameo lighting: A single foreground subject lit against a substantially uniform dark *background (3).*

camera: A device to convert an optical image, viewed by a lens system, into a permanent or temporary picture using photographic or electronic techniques.

camera cable: The interconnection between a *camera head* and its corresponding *camera control unit.* It is usually in the form of *triax,* but *optical fibre* cables are sometimes used at *outside broadcasts.*

camera card: A camera operator's *cue* card giving an individual camera's shots for a production.

Camera Control Unit (CCU): An *interface (1)* unit remote from and connected to a studio television *camera head.* It allows the camera output to be adjusted remotely and has provision for interfacing sound, communications, picture and *synchronization* signals.

camera duplicates: In *slide* production, photographing several originals of the same scene to avoid copying by duplication.

camera exposure sheets: Full instructions for every *frame (1, 2)* for use by a *rostrum camera* operator in *animation* photography.

camera head: The part of a studio television camera containing the lens and *CCD imagers* to frame, focus and image the scene. It is connected to a *camera control unit* via a camera cable.

camera light: See *basher.*

camera mount: A film or video camera support.

camera original: The film used in a motion picture camera to photograph the original scene.

camera ready artwork (CRA): For printing, *slides*, or *overhead projector* transparencies. Material ready to be photographed or *screened (4)*.

camera script: A script which includes details of shots, lighting and sound.

Canadian Association of Br oadcasters (CAB): A national industry association representing the majority of Canada 's private television stations, networks and speciality programming services. (W ebsite: www.cab-acr.ca).

Canadian Advanced Broadcasting Systems Committee (CABSC): A consultative group which operated from 1988 to 1990 to investigate when and how digital television services should be provided in Canada. (Hist.)

Canadian DTV Inc (CDTV): Canadian Digital Television (DTV) 'platform group' set up by the *CAB* to represent all sectors of industry interested in DTV conversion issues.

candela (cd): *SI unit* of luminous intensity, equal to $\frac{1}{60}$ the luminous intensity per square centimetre of the surface of a *black body* at the temperature of solidification of platinum.

Cannon connector: An earthed three-pin audio connector.

canoe: In the *quadruplex* video recording standard, the curved section of tape between the tape input and output guides. (Hist.)

cans: *Headphones* or *earpiece*. (Colloq.)

capacitance electronic disc (CED): A grooved capacitance *videodisc* system. (Hist.)

capacitor microphone: A *microphone* which converts the changing capacitance of a foil *diaphragm (2)* into an audio signal using an active pre-amplifier circuit usually contained in the body of the *microphone*.

capstan: The rotating shaft which provides constant speed tape motion in a magnetic recorder when held against the tape by a *pressure roller*.

caption: (1) A written or spoken title identifying or explaining the content of a pictorial image. **(2)** Term for artwork, lettering or *computer graphics* used for programmes.

card: A removable electronic circuit board.

cardioid microphone: A *directional microphone* with a heart-shaped *polar diagram*. See **Figure D.5**.

card reader: A device for reading *data* from magnetic or perforated cards. (Hist.)

Carousel: A rotary-*magazine (2)* gravity-feed *slide projector*, or the *magazine (2)* itself. (Trade name.)

carrel: A working booth for an individual student, usually fitted with *audio-visual* equipment such as an audio tape recorder, *slide projector* or video player.

carrier: (**1**) A constant alternating signal or *electromagnetic wave* which can convey information when it under goes *modulation* by another signal. Also called a *carrier signal* or *carrier wave*. (**2**) In USA *telecommunications*, an abbreviation for *common carrier*.

carrier signal: See *carrier*.

carrier-to-noise ratio (C/N or CNR): The ratio in *decibels* of the received *carrier* power to the *noise* power in a given bandwidth. It is used to assess the quality of *microwave* or *satellite (2)* communication links.

carrier wave: See *carrier*.

cart machine: Equipment at a radio or television *broadcast* station for storing and automatically playing a number of audio cartridges or *videotape* cassettes under *remote control*.

cartridge: (**1**) A container for a single *spool* of film or tape, feeding to a separate *spool*. (**2**) A container for a continuous loop of film or *magnetic tape*. (**3**) The *stylus (1)* holder and electromechanical *transducer* of a *gramophone* pick-up *head (1)*.

cassette: A plastic container enclosing both feed and take-up *spools* for film or *magnetic tape*.

cassette tape r ecorder (CTR): An audio recorder using magnetic tape *cassettes* as its recording medium.

catadioptric lens: An *objective lens*, usually of long *focal length*, which is part of a lens/mirror system combining transmitting and reflecting surfaces.

catch lights: Reflections of light in a performer 's eyes in a *close up* shot. Also called *eye lights*.

cathode ray tube (CRT): A glass *valve* narrower at one end and opening out into a luminescent picture display *screen (1)*. It converts a *scanning* beam of electrons into light of varying intensity and/or colour to form an image on the screen and was invented in 1897 by Karl Ferdinand Braun of the University of Strasbour g.

Cb: The (B–Y) *colour difference signal* reduced to 0.7 volts in *amplitude* and *bandwidth* limited to 2.75 MHz.

C-band: (**1**) The *electromagnetic spectrum* between 4 and 8 GHz with the 4 and 6 GHz bands being used for *satellite (2)* communications. Specifically, the 3.7 to 4.2 GHz communication band is used for *downlink* frequencies, with the 5.925 to 6.425 GHz band for the *uplink* (see **Appendix N**). (**2**) *Wavelengths* between 1530 nm and 1568 nm used in *optical fibre* communications systems.

CCD imager: The image sensor used in video cameras based on *charge coupled device* technology. It takes the form of a thin flat wafer ,

Figure C.1   CCD imager

comprising a mosaic of hundreds of thousands of light sensitive   *pixels*
which are coupled by integral CCDs to provide an analogue output
which follows the television system's *scanning* pattern. See **Figure C.1**
and **Figure D.4**.

Ceefax: The name of the   *BBC's teletext* service, derived from   'see
facts'.

cel: A transparent plastic sheet used for   *animation* or for *overhead
projector* transparencies. Sometimes spelt 'cell'.

cell: (1) An electrical battery or   *accumulator.* **(2)** The area covered by a
single aerial in a cellular radio system.   **(3)** A data transmission unit of
fixed length, e.g. *ATM* cells are 53 bytes long. **(4)** See *cel.*

central apparatus room (CAR): The area in a   *broadcast* studio centre
where the majority of the technical equipment is gathered together for
ease of interconnection and maintenance.

central processor unit (CPU): In a *computer*, the main data processing
component which executes the   *computer program* instructions.

**Centre Commun d' Études de Télédiffusion et Télécommunication
(CCETT):** The Centre for the study of   Television broadcasting and
Telecommunication in France. (Website: www.ccett.fr).

centre cut-out (CCO):   A way of viewing 16   :9 *aspect ratio (1)*
transmissions with an existing 4 :3 television and a   *set-top box.* A 4:3
patch is taken from the centre of the 16 :9 image, unless the broadcaster
has programmed the   *aspect ratio converter*   to take the patch from
another part of the image. There is considerable *cropping* of the image
at the sides.

Centronics interface: See *parallel port.*

ceramic cartridge: A *gramophone* pick-up *cartridge (3)* based on a *piezo-
electric* ceramic. (Hist.)

C-format: A 25mm (1 inch) *broadcast-quality helical-scan reel-to-reel* analogue *videotape recorder* system.

changeover: (1) Changing from one motion picture *projector* to another without interrupting continuity of presentation. **(2)** A type of electrical switch.

changeover cues: Visual indications, usually dots or circles, near the end of a *reel* of motion picture film to warn the operator to change from one *projector* to the other.

channel: (1) A discrete path carrying a specific signal from one user or device to another. **(2)** An allocated portion of a *frequency band* for transmission of a radio or television signal. **(3)** The circuit from a lighting control panel to the associated dimmer. **(4)** A television service, e.g. Channel 4.

channel coding: Converting a data signal carrying information into the most suitable form for the transmission medium being used. This may include the addition of *error detection and handling* data if the path is prone to *noise* or *interference* and an embedded clock signal to make the system *self-clocking*. See also *bi-phase mark* and *Manchester code*.

channel loading: A method of inserting film or *magnetic tape* into a player or *projector* by a lateral movement rather than by *lacing*.

Character And Pattern Telephone Access Information Network (CAPTAIN): Japanese *teletext* system.

character generator (CG): Equipment that produces the signals needed to format characters and symbols as *captions* and *titles* for a production.

characteristic curve: A graphical representation of the performance of a system, e.g. the relationship between *exposure* and *density* in photographic film. See **Figure C.2**.

characteristic impedance: The specified *impedance* of a cable which will result in no reflections in the cable if the *termination* at the end of the cable has the same value in *ohms.*, e.g. 75 *ohm* video cable with 75 *ohm* characteristic impedance must be terminated with 75 *ohms*.

character set: All the various characters available in a system. They are not always the same for *screen (1)*, printer and keyboard.

characters per second (cps): A measure of data transmission rate.

charge coupled device (CCD): An *integrated circuit* in which stored analogue information is moved between elements by char ge transfer from one element to the next. They were used in *Betacam* equipment to provide time compression, expansion or delay to the *component video* signals. In its main application a CCD forms an integral part of the *CCD imagers* used in video cameras.

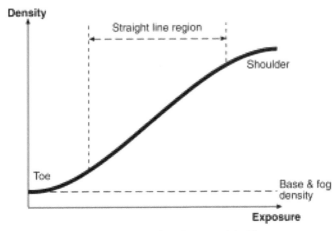

Figure C.2   Characteristic curve for photographic film

chase mode: *Synchronization* mode between two or more mechanical recorder/reproducers, where slaves follow the master without maximum interlock.

chatter technology: A process developed by *Quantel* for augmenting a *hard disk drive* system with *random access memory* to create fast multi-channel *random access* disks. (Hist.)

checkerboard cutting: (1) A method of assembling original 16mm film to allow prints to be made without visible joins. Alternate scenes of a reel of 16 mm film are assembled in separate   A   and B rolls with corresponding lengths of blank spacing. See *A/B roll* and **Figure C.3**. **(2)** A video editing mode where a computerized editing system records all the edits from the currently available playback tapes, leaving gaps to be filled later by material from other tapes.

checksum: Used to detect errors in data transmission. The data is grouped into *frames (4)* and an *algorithm* used to calculate a number depending on all the data in the *frame (4)*. This checksum is added to the end of the *frame (4)* before transmission. The received checksum is compared with one calculated in the *receiver* using the same *algorithm*. Any difference indicates one or more errors in the *frame (4)*.

cherry picker: Generic name for a lorry mounted hydraulic platform to raise cameras or *radio link* equipment.

chip: (1) An unmounted sub-miniature component, e.g. a chip capacitor . **(2)** An *integrated circuit*. (Colloq.)

chip camera: A video camera which uses one or more   *CCD imagers.*

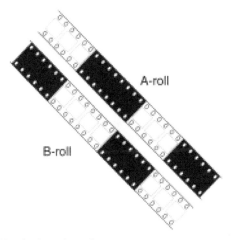

Figure C.3 Checkerboard cutting

chocolate block: A multi-pole electrical strip connector, often marked for sub-division. (Colloq.)

chop: An alternative term for a *hard cut.*

chroma: (1) The part of a *composite video (2)* signal carrying the colour information. Also called *chrominance.* **(2)** The degree of colour *saturation*, representing the purity of a colour , or how much white is added, i.e. low saturation colours are pastel shades.

chromakey: A video special ef fects technique in which areas of a high *saturation* colour (often blue) in a foreground scene are substituted by a picture from another source. *Keying* is used to electronically remove the coloured areas in the foreground and replace it with the *background (1)* image. Also called *colour separation overlay.*

chromatic aberration: A defect of a lens or optical system causing light of different colours to be focused in different planes. It is usually seen as coloured fringes around the components of an image.

chromaticity: The colour aspect of light which includes *hue* and *saturation*, but not *brightness.*

chromaticity diagram: A two-dimensional colour map used in assessing the design and *colorimetry* of colour television systems.

chroma timer compressed multiplex (CTCM): The basis of the colour recording in the *M II* video system. (Hist.)

chrome tape: Magnetic recording tape using chromium dioxide particles. It has higher *coercivity* and *remanence* than *ferric oxide tape*. Its full name is *chromium dioxide tape*.

chrominance: The part of a *composite video (2)* signal which conveys colour *hue* and *saturation (2)*, information, as distinct from *luminance (2)*. Also called *chroma (1)*.

chromium dioxide tape: See *chrome tape.*

chronophotograph: A camera for taking sequences of photographs in quick succession, devised by two German brothers, Max and Emile Skladanowsky in 1892.

cinching: (1) Uneven winding of *magnetic tape* on a *spool*, which may affect the audio or video quality on playback. (2) Tightening the turns in a loosely-wound roll of film.

cinch marks: Longitudinal abrasions on motion picture film or *videotape* resulting from the movement of one turn against another in a roll.

Cinema Digital Sound (CDS): A fully digital sound on film system introduced in 1990 for the film 'Dick Tracy' by the Optical Radiation Corporation, a division of Kodak. It only lasted two years. (Trade name.) (Hist.)

CinemaScope: A system of *anamorphic, widescreen* motion picture production and projection developed by Twentieth Century Fox which uses a horizontal compression/expansion factor of 2 :1. (Trade name.)

Cinématographe: Early film camera first used in 1895 by Louis Lumière.

cinematography: Recording and reproducing motion as a series of images on a strip of photographic film.

Cineon: A high *resolution* digital system for the production of special effects. (Trade name.)

Cinerama: A *widescreen* film system using three interlocked 35mm cameras and projectors with seven-track *magnetic film* audio. The first film, 'This is Cinerama', was shown in just two US cinemas but became the highest grossing film of 1952. The last screening in the original form was 'How the West was Won' in 1963. Cinerama folded in 1982. (Trade name.) (Hist.)

cine spool: A type of *magnetic tape spool* with a small keyed centre hole. It is usually made of plastic and 175mm (7 inches) or less in diameter.

circle of confusion: The amount of defocusing of a lens before the image becomes noticeably out of focus (see **Figure D.3**). For a television camera it approximates to the spacing of the CCD *image sensor* elements.

circle wipe: See *iris wipe* and **Figure W.1**.

Circuit (CCT): The interconnection of a number of electronic devices and components to perform an electrical function.

circuit breaker: A re-closable electromechanical safety device used to replace the function of a *fuse*. It may also be used as a switch.

circuiting: The direct transfer of *release prints* between cinemas rather than through an *exchange (1)*.

circuit switching: In a *wide area network,* establishing a route between two *stations* by making switched connections in all intermediate *exchanges (2)*. The route is maintained for the whole duration of the data transfer. Making a telephone call uses circuit switching.

circuit under test (CUT): The circuit being examined.

circular polarization: A *polarization* mode for an *electromagnetic wave* where the planes of the electric and magnetic components rotate around an axis along their direction of motion in a helix. There is one complete revolution in one wavelength of the wave. Two circular polarization modes are used, clockwise or *left-hand circular polarization* (LHCP) and anticlockwise or *right-hand circular polarization* (RHCP), depending on the direction of rotation (compare *linear polarization*).

Citizens Band (CB): A radio transmission band available to the public on specific frequencies and at low power to give a range of a few miles.

cladding: The transparent material surrounding the *core (3)* of an *optical fibre*. It has a lower *refractive index* than the *core (3)*, so optical or *infrared* energy is confined within the *core (3)*. See **Figure O.3**.

clamp: An electronic circuit which restores the DC level of a signal. Its main benefit is the removal of *mains hum* from the signal, which would otherwise produce *hum bars*.

clapper: The member of a film crew who operates the *clapper-board*.

clapper-board: In motion picture photography, a board with a hinged arm used to identify the scene being shot and to allow correct *synchronization* of picture and sound at the beginning or end of a scene. See **Figure C.4**.

Figure C.4 Clapper-board

class 1, class 2: In the safety classification of mains-powered equipment, Class 1 is earthed and Class 2 is double-insulated.

claw: A device in a motion picture camera or *projector* which engages and disengages the perforation holes to pull the film down and provide the *intermittent motion.*

clean feed (CF): (1) An audio monitoring mix giving a performer all signals but his own contribution. Called a *mix-minus* in the USA. **(2)** A video programme distribution feed taken prior to the addition of *captions.*

clean wind: Full-length fast-forward and rewind of *videotape* after *editing.*

clear vision: An enhanced *NTSC* television system developed in Japan to support the 16 :9 *widescreen* format and have improved performance over the *NTSC* system. (Hist.)

click track: A timing track consisting of recorded clicks, used for *dubbing* music with precise timing.

client-server network: See *server.*

clip: (1) A short film or video extract from a complete motion picture production. **(2)** A video shot or sequence stored in a *computer graphics* or *non-linear editing* system. **(3)** A selected part of a *computer graphic* to be moved or altered.

Clipnet: A high-capacity networking connection used by *Quantel.* It is based on *gigabit Ethernet*, using *NFS*, and it is an open standard. (Trade name.)

clipper: A form of *limiter.*

clipping: (1) *Waveform* distortion in sound or video caused by *peak amplitude* exceeding the available range. **(2)** In *computer graphics*, determining the parts of an image which lie outside a specified *clip (3)* boundary.

clock: A stable *oscillator* used for timing, especially in digital systems.

clock jitter: Unwanted random changes in the *phase (1)* of a *clock* signal.

clock track: (1) A time signal recorded on one track of an audio *magnetic tape*, to facilitate *editing* and *audio-visual* programming. **(2)** In digital magnetic recording, a timing track needed to recover data from the data tracks.

clock wipe: A *wipe* in the form of a rotating radius of a circle, like the hand on a clock. See **Figure W.1.**

clogging: The build-up of *oxide* and *binder* on a magnetic recording or playback *head (2)* causing *drop-outs.* Serious build-up can cause a *head clog.*

clone: In digital recording, an exact copy of an original digital recording.

closed captioning: US term for *subtitles*. The system has been expanded to permit transmission of other types of data, e.g. the *V-chip rating* data.

closed circuit television (CCTV): A television system where the camera output(s) are fed to local monitors and/or recorders and not *broadcast.*

closed GOP: An *MPEG-2 group of pictures* where the last frame of one GOP is a *P-frame* and does not use the first frame of the next (compare *open GOP*).

closed loop: A feedback or *servo system* in which a proportion of the output is looped back to be compared with the input requirement.

close-up (CU): A shot generally showing only the head of the subject.

cluster: An assembly of magnetic *heads (2)* for recording on *multi-track magnetic tape* or film.

C-mount: The standard threaded *lens mount* for 16mm film cameras and some video cameras.

C-multiplexed analogue component (C-MAC): An analogue component television transmission system. See *multiplexed analogue component*. (Hist.)

C/N: See *carrier-to-noise ratio.*

coax: See *coaxial cable.* (Colloq.)

coaxial cable: Cable having a central conductor within a cylindrical *screen (3)* layer. Often abbreviated to *coax.*

code: (1) To convert information into a different format, usually for efficient storage or transmission. See *encoder.* **(2)** In computers, the set of *instructions* forming a *computer program* when converted into *machine code.*

codebook excited linear prediction (CELP): A form of audio data compression using *psychoacoustic* coding. It is used in some low *bit rate* applications.

codec: A combination of an *encoder* and a *decoder.*

coded orthogonal frequency division multiplex (COFDM): The coded form of *OFDM* which adds error protection to the digital information being carried. This form is used in the *DVB* and *DAB* standards.

coder: See *encoder.*

coercivity: The magnetic energy needed to magnetize a material to a given level, usually its *saturation.*

coffin: A studio lamp housing containing many lamps and used to give shadowless illumination.

coherence: (1) In *computer graphics*, the property that adjacent *pixels* are likely to belong to the same display object, such as being inside or outside a given shape rather than being on the boundary of the shape. **(2)** In optics, the constant nature of the *wavelength* of a light source, particularly a *laser* source.

coincident pair: A *microphone* arrangement for *stereo* recording where two co-sited *microphones* are placed at 90 degrees to each other . The configuration may be a pair of *elements (1)* within a single *microphone* unit.

cold mirror: A *dichroic* surface reflecting visible light, but transmitting *infra-red* (heat) radiation so that it can be safely dissipated.

cold start: To *reboot* a system by turning of f and restoring the power (compare *warm start*).

collimated: A beam of light with all its rays parallel.

co-located: Two or more *satellites (2)* sharing the same *geostationary Earth orbit* assignment.

Colorama: Coloured paper available in rolls for studio backings, etc. (Trade name.)

coloration: Distortion caused by the addition of unwanted harmonics to an audio signal.

colorimetry: The study and application of colour matching accuracy using three *primary colours*.

colour analyser: Equipment used to view colour film negatives as a positive image by *closed circuit television* for assessment of the printing levels required.

colour balance: The appearance of a colour image considered in terms of its three *primary colour* components.

colour bars: A video test signal used to confirm the correct operation of a video system. It provides vertical stripes of white, yellow , *cyan*, green, *magenta*, red, blue and black across the picture. There are several different types of colour bar signals. See **Appendix Z**.

colour black: A video signal without picture information. It contains only *sync pulses*, *colour burst* and *black level*. It is used as a reference signal for timing purposes in a television studio installation and may also be called *black and burst* , *black and syncs* or *black burst*. See **Figure C.5**.

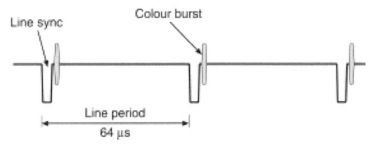

Figure C.5 Colour black line waveform

colour burst: A few reference cycles (8 to 10 in *NTSC*, 9 to 11 in *PAL*) of *subcarrier* transmitted during the *back porch* period of a *composite (2)* colour television signal. See **Appendix K.**

colour correction: Changing the *colour balance* or other characteristics of an image to improve the subjective image quality .

colour difference signals: The signals (B –Y) and (R –Y), derived from *RGB*, from which the *chrominance* signal in a *PAL* or *NTSC* encoder is made. With different *weighting* they are used to generate the *Cb* and *Cr* colour signals used in analogue and digital *component video* systems.

colour dub: A facility provided on some *video cassette recorders* which provides improved *dubbing* performance.

colour filter: A transparent material which selectively absorbs specific regions of the *visible spectrum.*

colourization: The electronic addition of colour to a *videotape* transfer of a *B&W* motion picture.

colour mobile contr ol room (CMCR): A vehicle used on an *outside broadcast.*

colour reversal intermediate (film) (CRI): A duplicate colour negative processed by reversal. (Hist.)

colour separations: A set of *monochrome* photographic images recording colour information in limited spectral ranges, usually red, green and blue.

colour separation overlay (CSO): A video *keying* technique. See *chromakey.*

colour temperature: A measurement of the colour quality of light. It corresponds to the temperature on the *Kelvin* scale at which the visible radiation of a *black body* matches that of the source in question. The higher the colour temperature, the more blue the light (*daylight*), the lower the colour temperature, the more orange (*artificial light*). Typical figures are 6500K for *daylight* and 3000K for *artificial light*. See also *mired* and **Table C.1.**

colour under: A method of recording analogue *composite video (2)* on videotape. *Chroma* is recorded on a lower *carrier* frequency than the *luminance* signal. The system is used in domestic tape formats such as *VHS* and in the *U-matic* format.

colour wash: (1) A facility for the overall colouring of individual *screens (1)* in a *videowall.* **(2)** Adding an overall colour to *computer graphics.*

column: An enclosure containing a vertical array of *loudspeakers.*

coma: A lens *aberration* causing a point off the *optical axis* to be imaged as an asymmetric patch of light with a tail like a comet. It can be improved by *stopping down* the lens.

comb filter: A *filter (3)* with notches in response tuned to a fundamental frequency and its *harmonics* but leaving intermediate frequencies

Table C.1 Colour temperatures of various light sources

Source	Colour Temperature	
	Kelvin	Mired
Candlelight	1900 K	526
Dawn or dusk	2000–2500 K	500–400
Room lighting (40–200 W)	2600–2900 K	385–345
Tungsten halogen studio lamps	3200 K	312
Photo flood lamps	3400 K	294
White fluorescent	3500 K	286
Late afternoon sunlight	4500 K	222
Summer sunlight	4800–5000 K	208–200
Average sunlight	5500 K	182
Xenon arc	6400 K	156
Electronic flash	6500 K	154
Overcast sky	7000 K	143
Summer shade	7100–8000 K	141–125
Hazy blue sky	9000 K	111
Clear blue sky	up to 30 000 K	down to 33

unaffected. Its *frequency response* therefore resembles the teeth on a comb. Commonly used in *composite video (2)* decoding to separate *luminance* and *chrominance*.

combined print: See the preferred term *married print*.

Comité Consultatif International de Radiodiffusion (CCIR): International Radio Consultative Committee of the *ITU*. It was replaced in 1994 by ITU-R (see *International Telecommunications Union*). (Hist.)

Comité Consultatif International du Télégraphe et du Téléphone (CCITT): *ITU* Consultative Committee for International Telegraphy and Telephones. It was replaced in 1994 by ITU-T (see *International Telecommunications Union*). (Hist.)

Comité Européenne de Normalization Electrotechnique (CENELEC): European committee for electrotechnical standardization. (W ebsite: www.cenelec.be).

commag: Composite magnetic print. A motion picture film carrying a magnetic sound-track alongside the picture image. (Hist.)

commentary track: A track on which a narrative, commentary or *voice over* is recorded.

Commission Internationale de l 'Éclairage (CIE): The international commission on illumination, based in Vienna, Austria. (Website: www.cie.co.at/cie/home.html).

Commission Nationale de Communications et de la Libert é (CNCL): The French *broadcast* regulatory body.

common carrier: In the USA, an or ganization which supplies *telecommunications* services to the public. The equivalent European term is *PTT (Post, Telegraph and Telephone)*.

Common Image Format (CIF): Proposed universal video acquisition and production format based on 24 *frames per second* with *progressive scanning*. It has 1920 *pixels* by 1080 lines and a *widescreen aspect ratio (1)*. Also called *HD-CIF*.

Common Intermediate Format (CIF): A digital video picture format with 352 *pixels* by 288 lines at up to 30 *frames per second*, used for *video conferencing* applications. The frame rate can vary from between 1 and 30 *frames per second*, depending on the data rate available. See also *Quarter Common Intermediate Format* .

common mode: In-phase signals in a *balanced* circuit.

common mode rejection: The ability of *balanced* or *differential* inputs to reject in-phase signals.

common mode rejection ratio (CMRR): The ratio of the *common mode interference* on the input to a system and that on its output.

Commonwealth Broadcasting Association (CBA): An association funded by subscription from members of the major public service broadcasters of the Commonwealth. Its stated aim is 'Working for Quality Broadcasting throughout the Commonwealth'. (Website: www.cba.org.uk).

Communication Concentrator Unit (CCU): A *computer* device which handles most of the work of a multi-user *computer* system by controlling the display of information on the *terminals* connected to it. It also interprets *terminal* commands and may carry out word processing functions.

Community Antenna Television (CATV): Television signals distributed by cable from a central *aerial*. The acronym is also applied to *cable television*.

comopt: Composite optical print. A motion picture film carrying a photographic soundtrack alongside the picture image. See **Appendix G**.

compact cassette: The standard audio or data cassette using 3.38mm ($\frac{1}{8}$ inch) wide *magnetic tape*. This is not to be confused with the miniature cassettes used in dictation or answering machines. (Registered design.)

compact disc (CD): A disc of 12cm diameter or less used to record digital audio, video or data and read by an optical *laser* system.

compact disc – digital audio (CD-DA): A *CD* containing digital audio data, defined in the *Red Book*.

compact disc – erasable (CD-E): See *compact disc – read write*.

compact disc – extra (CD-Extra): An enhanced multi-session music *CD* which contain audio and data. It is defined in the *Blue Book (2)*.

compact disc – interactive (CD-i): An *interactive* system incorporating *compact disc-video* recordings, developed by Philips in 1987 for the consumer market. It is defined in the *Green Book*.

compact disc – read only memory (CD-ROM): A *CD* containing digital data. It is defined in the *Yellow Book*.

compact disc – read write (CD-R W): A re-writable form of *CD*. Originally called *CD-E*.

compact disc – recordable (CD-R): A blank *CD* onto which data can be recorded using a *CD* writer.

compact disc – video (CD-V): An obsolete *disc* format containing analogue video and sound. It is not the same as *Video CD*. (Hist.)

compact iodide, daylight (CID): A light source of 5500K. (T rade name.)

Compact Source Iodide (CSI): A particular form of *metal-halide lamp* having a *colour temperature* of about 4300K. (T rade name.)

compander: Combined audio *compressor* and *expander*.

comparator: (1) An electronic circuit whose output indicates the result of comparing the voltages existing at its two inputs. (2) A *computer graphics* device that compares the proximity of a *cursor* to the *vector (2)* currently being drawn.

compensator: The lens or group in a *telephoto lens* which moves to maintain focus (compare *variator*).

compiler: A *computer program* which converts *source code* (written in a *high level language*) into executable *object code* for subsequent execution.

complementary colours: In light, two colours which produce white light when added together, thus red and *cyan* (blue-green) are complementary colours.

component analogue video (CAV): See *analogue component*.

component video: A set of signals, each of which represents a portion of the information needed to generate a full colour video image. The signals are usually distributed separately as *RGB* or *YCbCr* (sometimes inaccurately referred to as *YUV*). Sometimes a fourth signal is added either carrying *synchronization* information (the S signal) or *keying* information (the K signal).

composite: (1) A photographic *montage* of images, as in *multi-screen.* **(2)** Several images on a single *slide mount.*

composite print: See the preferred term *married print.*

composite video: (1) A television signal with both picture and *synchronization* information. This was the original meaning as applied to *monochrome* television systems. (Hist.) **(2)** An *encoded (3)* television signal containing picture, *blanking*, colour and *synchronization* information. This is the meaning when referring to colour television systems.

compositing: Combining images with *keying*, graphics and associated audio to produce a complete graphic sequence.

compound lens: A lens system with several lenses combined to correct or reduce *aberrations.*

compound table: A device to provide X, Y and rotary movement used with a *rostrum camera* in *animation* and in the production of streak and other effects in *slides.*

Compressed Serial Digital Interface (CSDI): Proprietary interconnection standard developed by Panasonic for carrying digitally compressed video and uncompressed audio signals using *coaxial cable* and *BNC* connectors. Signals can be transferred four times faster than real-time. The *interface (1)* is electrically compatible with the *serial digital interface* standard.

Compressed Time Division Multiplex (CTDM): The basis of the colour recording in the *Betacam* and *Betacam SP* video system.

compression: (1) In audio, reducing the *dynamic range* of a sound source. **(2)** In digital video or audio, using an *algorithm* to reduce the amount of data needed to describe the video images and audio information. It is more correctly called data compression to distinguish it from (1).

compressionist: An operator who regulates and adjusts the *compression (2)* process to give better results than would be achieved by an automated system.

compressor: (1) A device restricting the *dynamic range* of audio programme material to an adjustable narrower range. **(2)** A device able to compress digital data to fit within an available *channel (1) bandwidth.*

computer: An electronic device designed to perform a wide range of tasks, using some form of flexible *programming* technique. See also *microprocessor.*

computer aided design (CAD): Using a *computer* to generate or originate graphic material, often in the form of technical drawings, e.g. for architectural or engineering use.

computer graphics: Images originated or *animated* by *computer.*

computer graphics imaging (CGI): Generating images using sophisticated *computer* processing techniques.

computer language: A specific vocabulary and syntax used to communicate with and operate a *computer*. See also *high level language*.

computer output to microfilm (COM): Recording information as images on photographic film which can then be stored and read using a special reader.

Computer Users Tape System (CUTS): Standard used for recording data on to *compact cassettes*. (Hist.)

computer virus: A small *program* which can enter a *computer* via a *modem*, *CD-ROM* or *floppy disk* and cause it to malfunction. The code can rapidly replicate by attaching itself to otherwise benign *programs* or information and entering other *computers* in a cascade effect.

concatenated code: In digital recording, linked blocks of coded information used for *error correction*.

concatenation: Linking together several data *compression (2)* systems which involves decompressing and recompressing the signal.

conceal/reveal: A video *transition* in which one scene appears to slide across another: in *conceal* the new scene obscures the previous one, in *reveal* the previous scene moves to display the new scene beneath (compare *push, scroll*).

condenser lens: A lens, lens system or lens/reflector combination used to concentrate light from a source into a defined beam, as in a *projector*.

condenser microphone: See *capacitor microphone*.

conditional access: In digital broadcasting, controlling access to a particular channel or service by *scrambling* the digital television signal so that it can only be descrambled by a transmitted code or a pre-paid access card. It can be used for *pay TV, pay per view* or to restrict the coverage area for a service.

conditional access table (CAT): A data structure carried in an *MPEG-2* digital *multiplex (2)* whenever any programme in the *multiplex (2)* is *encrypted*. It contains a list of the *PIDs* carrying the *Entitlement Management Messages (EMMs)* for the *conditional access* system(s) in use. It is part of the *MPEG-2 Programme Specific Information (PSI)*. See **Appendix O**.

cones: The cells in the *retina* of the human eye which are sensitive to colour (compare *rods*). There are three types of cone. See **Figure C.6** and also *human eye* and **Figure H.6**.

conforming: The assembly of picture and sound components in film or *videotape* to match the defined *editing* continuity, often using an *edit decision list*.

console: A control desk for lighting or sound.

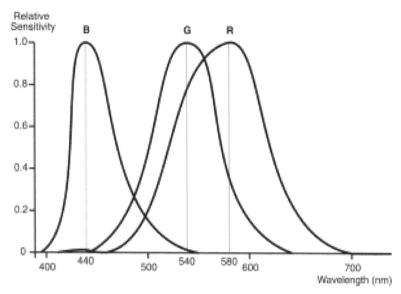

Figure C.6   Relative colour responses of cone cells

constant angular velocity (CAV): An optical disc storage system in which data is recorded in a series of concentric rings and the disc is maintained at a constant rotational speed, similar to that used in a *floppy disk* system.

constant bit rate (CBR):   A data transfer needing a fixed bit rate, e.g. digitized audio and video which are delay sensitive such as *compact disc – digital audio*.

constant current: A device or circuit capable of maintaining a steady current in spite of changing voltage or varying *load (2)*.

constant linear velocity (CLV): An optical disc recording system which gives a constant track recording density but variable rotational speed, e.g. in a *compact disc* system.

constant voltage transformer: A *transformer* designed to supply a constant output voltage in spite of any changes in input voltage.

constellation: (1) The pattern of spots produced by *QAM* which would be seen on an *oscilloscope* displaying the two modulating signals as X and Y inputs. **(2)** A group of more than one *telecommunications satellite (2)* in a single orbital position.

Consumer Electronics Manufacturers Association (CEMA): A division of the *EIA* that represents manufacturers of consumer electronics (USA). (Website: www.cemacity.org).

contact bounce: A mechanical fault in a pair of electrical contacts.

contact microphone: A *microphone* designed to respond to sound vibrations transmitted other than through air .

contact printing: Still or motion picture printing in which the processed film to be copied is held in contact with the *raw stock* at the time of *exposure.*

continuity: (1) The correct sequence and matching of all aspects of action and setting between successive scenes in a motion picture or television production. (2) The documents prepared to ensure such results. (3) Link announcements between sequences or between *broadcast* programmes.

continuity still: A photograph of a scene taken to record details of setting, costumes, etc., to ensure matching in subsequent shots.

continuous animation: An *interactive* system of *animation* as opposed to pre-programmed *graphics.*

continuous cassette: An audio *compact cassette* containing an endless loop of tape.

continuous printing: Motion picture printing in which the processed film to be copied and the *raw stock* move continuously past the point of *exposure* (compare *step printing*).

continuous projection: (1) Motion picture projection in which the film moves continuously past the illuminated *aperture (1)*, the necessary *intermittent motion* being obtained by optical means. (2) Projection of film as a continuous loop so that it may be presented time after time without a break. A normal *intermittent motion projector* may be used in conjunction with a loop *platter* or cassette.

continuous tone masks: See *soft-edged masks.*

contour correction: See *aperture correction.*

contouring: An unwanted distortion of a digital video picture when the values used to describe each *pixel* are too coarse. The effect is similar to *posterization.*

contrast: The range of light intensity between light and dark areas in a scene or on a film or video image.

contrast control: In television picture display equipment, the contrast control sets the perceived intensity of light areas of the picture.

control characters: Non-printing *ASCII* characters used for control functions. They are not displayed on a *visual display unit*.

control key (CTRL): This key has to be held down to obtain *control characters* on a keyboard.

control room: See *gallery.*

control track (CTL): (1) In an audio tape, a track carrying instructions on operations to be performed during running, e.g. mixing, lighting, projection, etc. **(2)** In *videotape* recording, a *track (1)* used to provide frame pulses to allow correct *lock-up* during replay. It may also be used in *videotape editing*. **(3)** In motion picture films with multiple (magnetic) sound tracks, a track controlling the distribution of sound to the various *loudspeaker* systems.

convergence: (1) The adjustment of electron or optical beams for exact alignment of the three *primary colour* images in a *CRT*-based colour video display or *projector*. The images should superimpose exactly to avoid the appearance of coloured fringes around objects in the displayed image (compare *registration (2)*). **(2)** A term used to describe the amalgamation of different technological areas, for example computing with broadcast to enable desktop editing and server -based transmission systems to be developed.

cookie: (1) An irregular shaped light shield used to create shadow ef fects such as dappling or shadow patterns. Also called *cucalorus* or *kukaloris* or *kukie*. See also *branchaloris*. **(2)** A small set of *computer* instructions downloaded to the user 's *computer* from a *website* being accessed via the *Internet*. It is designed to speed up subsequent access to the same *website*.

Co-ordinated Universal Time (CUT): See *Universal Time Co-ordinated*.

copier: A device for duplicating images or re-recording sound.

core: (1) A cylinder, usually of moulded plastic, on which a roll of film or tape is wound. **(2)** The centre of an inductor on which the turns of wire are wound. **(3)** The light-transmitting material at the centre of an *optical fibre*. See **Figure O.3**.

coring: A video noise gating process in which video below a pre-determined luminance threshold is replaced by *black level*. See also *noise slice*.

cornea: The coloured front part of the *human eye*. See **Figure H.6**.

corner pinning: Manipulating an image after fixing one point so that the movement occurs around the fixed point.

corner wipe: A *wipe* in the form of a box from one corner . See **Figure W.1**.

co-sited sampling: A *sampling* technique applied to *colour difference component video* signals where *sampling* takes place at a submultiple of the *luminance sampling frequency.* This means that a *colour difference sample* is taken at the same instant as a *luminance sample.*

cosmic rays: Subatomic particles of great energy originating in outer space and reaching the Earth from all directions.

coulomb (C): *SI unit* of electric charge, named after the eighteenth century French physicist Charles A. de Coulomb. It is the quantity of electricity transferred by a current of one *ampere* each second.

counter: A mechanical, electromechanical or electronic device for counting operations. In motion picture practice, a counter measures the number of *frames (1)* along the length of the film, expressing the result in *frames (1)* only, or in feet and *frames (1)*.

country code (CC): See **Appendix P.**

coupling: Connecting two circuits or systems together. Circuits connected with a wire are directly coupled; circuits connected through a capacitor or a transformer are indirectly (or *AC*) coupled.

Cox Box: A colour *caption* synthesizer, also used as a generic term for *matte* generators. (Trade name.) (Hist.)

Cr: The (R–Y) *colour difference signal* reduced to 0.7 volts in *amplitude* and *bandwidth* limited to 2.75 MHz.

crab: To move a film or video camera or *microphone* sideways, often on a *dolly.* Not to be confused with *pan.* Also called *trucking* in the USA.

crane: (1) A large wheeled *camera mount* with a *boom* arm to lift and support the camera and operator above the *set* or *location* during shooting. **(2)** To move a film or video camera vertically .

crash: The failure of a *computer* to continue to execute the intended instruction sequence. It can be caused by *interference,* fleeting power failures, *software* incompatibilities or other *software bugs.*

crash recording: Switching a *videotape recorder* direct from play-back mode to record. The new recording may not be in correct *synchronization* with the earlier one.

crawling title: A line of *titles* or *captions* moving horizontally across the *screen (1)*.

credits: Acknowledgement given in publication, especially the *titles* at the beginning or end of film or television productions, listing the cast, technicians and organizations concerned.

creeping sync: In film recording, a progressive error of *synchronization* between picture and sound track, and the steps taken for its correction.

creeping titles: See *rolling titles.*

crispener: An electronic device for subjectively enhancing video picture image edges.

critical-damping: *Damping* in which a steady state is reached in the minimum time with one oscillatory cycle. See **Figure D.1**.

cropping: Cutting off the top and/or bottom of a projected picture to present a wider *aspect ratio (1)*.

cross colour: *Brightness* information incorrectly decoded as colour on a *composite video (2)* picture. It is particularly apparent in areas of fine picture *detail*, such as checked or striped shirts and jackets.

cross fade: A *mix transition* between two audio or video signals or between two *slide projectors.*

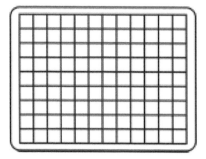

Figure C.7   Cross hatch picture display

cross hatch: A test signal for adjusting the *convergence* and geometry of television picture display equipment, comprising a grid of interconnected horizontal and vertical lines. Also called *grille*. See **Figure C.7**.

cross hairs: The engraved cross at the centre of the film *frame (1)* on a film camera *viewfinder (2)*.

cross light: See *kicker*.

cross luminance: See *dot crawl*.

cross mod: Cross-modulation test. In motion picture practice, a method for determining the optimum printing requirement for a photographic variable area sound record.

crossover distortion: A form of distortion in some *amplifier* circuits arising from the changeover from one output device to the other .

crossover network: A *filter (3)* network dividing an audio channel between two or more *loudspeakers*.

crosstalk: Unwanted breakthrough of one electrical signal onto another .

cross wipe: A *wipe* in the form of a cross. See **Figure W.1**.

crowbar: A power supply protection system which shorts the supply rails if over-voltage occurs.

crow's foot: A notched holder to prevent the legs of a *tripod* slipping. See also *spreader*.

crystal: See *quartz crystal*.

crystal sync: A method of *synchronizing* an audio *magnetic tape* recorder to a motion picture camera.

cucalorus: See *cookie (1)*.

cue: **(1)** A command or signal for a previously specified event to take place. For example, action or speech to commence or for a device to carry out the next item of its sequence.   **(2)** To prepare an *audio disc* ready to replay a track or a *videotape recorder* to replay a shot, scene or whole programme.

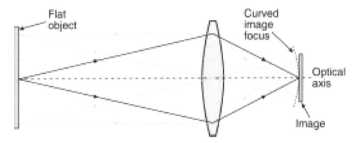

Figure C.8 Curvature of field for a lens

cue/review: A control on an audio tape recorder , to keep the replay *head (2)* in contact during fast winding.

cue dots: Visible signals to indicate the end of a programme section. In film, usually white circles in the top right-hand corner of the *frame (1)* for *projector* run-up and change-over . In television, usually a white square in the top right or left corner of the *screen (1)*.

cue light: The red lamp on top of a video camera. Also called a *tally light*, particularly in the USA.

cue tone: An *audio frequency* of specified duration recorded on a *cue track* to provide the indication of the *prompt* or *cue*.

cue track: (1) A control track used for *tape-slide synchronization* or control purposes on audio tape recorders and *cart machines*. **(2)** A secondary audio track on a *videotape* recording, available for electronic labelling or for a secondary programme channel.

cukuloris: See *cookie (1)*.

Current Bit Rate (CBR): An *ATM* service class providing a virtual circuit with a fixed data transfer rate. See also *available bit rate (ABR)* , *uncommitted bit rate (UBR)* and *variable bit rate (VBR)* .

current loop: In data transmission, a *serial communication* system where current flowing represents a *mark* and no current represents a *space*.

cursor: The character indicating the currently active position on a *computer graphics* display screen.

curvature of field: A lens *aberration* in which the sharpest image of a plane object is formed in a curved surface rather than a plane. It can be improved by *stopping down* the lens. See **Figure C.8**.

curved-field lens: A projection lens designed to focus a curved surface rather than a plane. It can be used to project cardboard mounted *slides* which have 'popped'.

cushioning: Smoothing sudden changes in motion in an *animation* to make them look more realistic.

cut: **(1)** A command given by a film director to stop the performance of a scene and its photography. **(2)** In *editing*, an instantaneous change from one scene to another . **(3)** In vision mixing, an instantaneous change between two video pictures, also called a *take (2)*. **(4)** In *slide* projection, a rapid change from one image to the next produced by lamp control. In order of increasing speed they may be termed Cut, Fast Cut and Hard Cut. The very fast *snap* is produced by *shutters (3)*.

cut and paste: **(1)** In *computer graphics*, moving an item to a new location on a display . **(2)** A word-processing term for combining blocks of standard text with variable inserts to create personalized documents, such as mailshots.

cutaway: A non-critical shot used to break or link principal action in scenes.

cut-out: **(1)** An electrical protective device to break a circuit on *overload (2)* or overheat. **(2)** In *animation*, small pieces of artwork cut to shape for movement frame by frame during photography . **(3)** A flat composition cut to outline, used in studio set construction.

cutter: A long thin *flag (1)* to control part of the light beam from a *luminaire*.

cutting copy: See *work print*.

cutting frames: Extra *frames (1, 2)* at the start and finish of an *animation* scene to give latitude in *editing*.

CV-One: A *videotape recorder* format from Hitachi recording video on 6.75 mm ($\frac{1}{4}$ inch) tape with separate *luminance* and colour tracks. (Hist.)

C-weighting: A *frequency response* weighting network used for measuring some types of *noise*. The network is defined in *IEC* Recommendation 651.

cyan: The *complementary colour* to red.

cyc: See *cyclorama*. (Colloq.)

cycles: **(1)** In animation, a set of artwork for an action which repeats itself after a certain number of *frames (1, 2)*. **(2)** Recurring periods of events. In electrical terminology, the *frequency* of a signal is expressed as the number of repetitions in each second, or cycles per second (see *Hertz*).

cyc lights: See *groundrow lights*.

cycle-time: See *dissolve cycle time*.

cyclic redundancy check (CRC): A method of verifying digital data.

cyclorama: A smooth curtain or back cloth suspended around the periphery of a studio or stage.

cylinder: A virtual tube passing through all the *surfaces* of a *hard disk drive*. All the tracks corresponding to one position of the head stack. See **Figure H.2**.

C-zero: A *compact cassette* without any *magnetic tape*.

D

D1: *DVTR* format, developed by Sony , which records 8-bit *component video* signals to the *ITU-R BT.601* standard on a 19 mm (¾ inch) tape cassette with four digital audio tracks.

D16: Recording format for digital film images based on the *D1* format. Three film *frames (1)* are recorded every two seconds. Full-motion replay is possible at ×16 tape speed.

D2: *DVTR* format, developed by Ampex, which records a *composite video (2)* signal sampled at four times the *subcarrier* frequency on 19mm (¾ inch) *D1* type tape cassettes with four digital audio tracks. (Not to be confused with *D2-MAC*.)

D2-multiplexed analogue component (D2-MAC): An analogue component television transmission system. (Not to be confused with the *D-2* digital *VTR* standards.) See *multiplexed analogue component*. (Hist.)

D3: *DVTR* format, developed by Panasonic, which records an 8-bit *composite video (2)* signal sampled at four times *subcarrier* frequency on a 12.5mm (½ inch) tape cassette with four digital audio tracks.

D4: There is no D4 format since it would translate as 'death' in Japanese.

D5: *DVTR* format developed by Panasonic using a 12.5mm (½ inch) *D3* type tape cassettes, but recording 10-bit *component video* signals to the *ITU-R BT.601* standard and four digital audio tracks. Defined in standard SMPTE 279M (see **Appendix U**). See also *HD-D5*.

D6: *DVTR* format, developed by Philips, which records uncompressed *high definition television* digital *component video* at 1.2 Gbit/s on a 19mm (¾ inch) *D1* type tape cassette. Defined in standard SMPTE 277M (see **Appendix U**).

D7: Format used in the *DVCPRO DVTR* system, developed by Panasonic, which records a *DV compressed* digital *component video* signal at 25 Mbit/s on a 6.75 mm (¼ inch) tape cassette using *metal particle* tape. Defined in standard SMPTE 306M (see **Appendix U**).

D-8: See *Digital 8*. Note this is not one of the SMPTE D series of digital tape formats.

D9: Format used in the *Digital-S DVTR* system, developed by JVC, which records a *DV compressed* digital *component video* signal at 50 Mbit/s on a 12.5mm (½ inch) *VHS* type tape cassette. Defined in standard SMPTE 316M (see **Appendix U**).

D9-HD: *DVTR* format for *high definition television* based on *D9* and recording *compressed* digital *component video* on 12.5mm (½ inch) cassette tape at 100 Mbit/s.

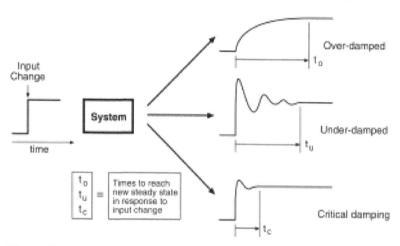

Figure D.1   Damping

dailies: See *rush prints*.

damping: Controlling the response of a system to input changes, for example a *servo system*. *Critical damping* occurs when a steady state is reached in the shortest time after only one cycle of oscillation.   *Over-damping* takes a much longer time.   *Under-damping* takes longer than one cycle and the system may be unstable and suf fer from *hunting (1)*. See **Figure D.1**.

damping factor: The nominal *load (2) impedance* of an *amplifier* divided by its actual output *impedance*. A measure of the *damping* effect on the connected *loudspeaker*.

damping wires: Thin horizontal wires in an *aperture grille cathode ray tube* to support the grille. They leave a faint shadow across the displayed image.

dark fibre: *Optical fibre* which has been installed by one company for use by another who provide their own interfacing equipment for data transmission along the fibre.

dashpot: A mechanical *damping* device which may consist of a body moving in a viscous medium.

data: *Alphanumeric* or other information fed to or from a   *computer* or digital device.

database: A store of related information and the *software* which allows the information to be or ganized and accessed by a   *computer*.

database management system (DBMS): An *application program* making data storage, retrieval and updating as easy as possible for the user of a *database*.

datagram: A data *frame (4)* used in *Internet protocol.*

data tablet: The flat working surface used in *computer graphics* for the input of information, with a *stylus* for drawing or a *puck* for digitizing. See also *graphic tablet.*

daylight: Illumination by natural sun and sky or by a source of equivalent *colour temperature,* generally between 5000K and 6000K. Colour film for daylight *exposure* is normally balanced to 5500K, 'average' sunlight.

daylight projection: *Screening (4)* of images in high ambient light. Requires a high-gain *screen (1)* with a narrow viewing angle, but *back projection* may be preferred.

daylight screen: A projection surface of high reflectivity (and hence narrow angle of view) for use in high ambient light conditions.

dBA: *Decibels* measured with *A-weighting.* This scale is commonly used for measuring ambient acoustic or electrical *noise.*

dBk: Power level in *decibels* relative to 1 kilowatt (i.e. 0 dBk = 1 kilowatt).

dBm: Power level in *decibels* relative to 1 milliwatt in a 600 *ohm* circuit (i.e. 0 dBm = 1 milliwatt in a 600 *ohm* circuit).

dBmV: Voltage ratio in *decibels* relative to 1 millivolt at 75 *ohms.*

dBu: Voltage ratio in *decibels* relative to 0.775 volt.

dBV: Voltage ratio in *decibels* relative to 1 volt.

dBW: Power ratio in *decibels* relative to 1 watt into an 8 *ohm load (2)* when used in audio *amplifier* specifications. The term is often used when considering *satellite (2)* reception.

dbx: An audio *noise reduction* system that encodes a signal by *compressing* with a 2:1 ratio. After storage or transmission, the signal is expanded in a 1:2 ratio to restore its original *dynamic range.* The name refers to its developer, David Blackmer. (Trade name.)

D-channel: The 16 kbit/s or 64 kbit/s digital control and signalling channel in *Integrated Services Digital Network* systems.

dead black-out (DBO): Total and usually sudden darkness on stage or in a studio.

debugger: A *computer program* which helps locate logic errors in a user program.

debugging: Locating and correcting errors in an electronic system or *computer program.*

decibel (dB): A unit usually used to give a measure of relative levels of intensity, power, or voltage, although it may refer to any other quantity . It is equal to one tenth of a *bel.*

deck: (1) The turntable and drive for an *audio disc.* **(2)** For *magnetic tape,* a tape *transport* mechanism with the *heads (2)* and associated electronics, but generally not including any *power amplifiers.*

decoder: A device used to recover a signal in one format from a processed version in a dif ferent format, e.g. a *PAL* decoder recovers *component video* signals from a PAL *composite video signal* (compare *encoder*).

decoding: The process of restoring coded data to its original form.

Decoding Time Stamp (DTS): In *MPEG-2*, an indicator of the decoding time for a compressed *MPEG-2* picture *frame (2)* or audio *frame (5)* (compare *presentation time stamp*). See **Appendix O.**

dedicated: Equipment designed to do one task, e.g. a purpose-built word processor, as opposed to a general purpose *computer* running a word processor *program.*

de-emphasis: Restoring the original *frequency response* of a signal after transmission or recording which involved *pre-emphasis.*

de-esser: A *limiter/compressor* and *filter (3)* combination which controls excessive sibilance ('ss' sounds) in dialogue recordings.

default value: The value which will be assumed in the absence of an alternate instruction. For example, in word-processing, the number of lines per page could be assumed as 66 in default of entering a figure.

definition: A description of the sharpness or clarity of a television picture. High definition pictures portray a lot of *detail,* while low definition pictures look *soft* and less clear. See also *resolution.*

deflection yoke: See *yoke.*

defragmentation: In a disk-based storage system, rearranging the data held in disk *sectors* so that *files* are stored on consecutive *sectors* in adjacent *tracks (3).*

degauss: To remove residual magnetism, particularly that built up in a *cathode ray tube* display.

degausser: A device for reducing residual magnetism, for example in tape *heads (2).*

delay line: A device for delaying information using a *transmission line.* Sometimes the delay is only for providing a short-term store.

delay PAL: A PAL decoding system where the colour is decoded using a *delay line* so that any phase errors are only visible as a slight *desaturation* (compare *simple PAL*).

delta gun tube: A type of colour *cathode ray tube* with its three *electron guns* in a triangular formation rather than in a line as found in an *in-line tube.* See **Figure D.2.**

delta modulation: A form of *modulation* where changes in the level or value of the signal to be carried are used rather than the actual level or value of the signal.

demodulation: Recovering the wanted signal from a *modulated carrier wave.*

demodulator: A device for recovering the wanted signal from a *modulated carrier wave.*

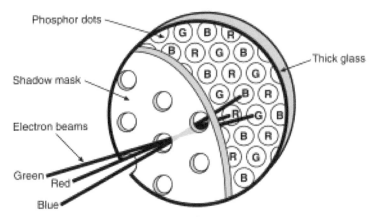

Figure D.2   Delta gun cathode ray tube

density: In photography, a measure of light attenuation, for example of a photographic image. It is the reciprocal of *transmission (2)*.

densitometer: An instrument for measuring the amount of light reflected or transmitted by an object.

de-orbit: To remove a *satellite (2)* from its *geostationary Earth orbit* assignment.

Department for Culture, Media and Sport (DCMS): UK Government department established in July 1997 and responsible for arts, broadcasting, the press, museums and galleries, libraries, sport and recreation, historic buildings and ancient monuments, tourism, the music industry , and the National Lottery. (Website: www.culture.gov.uk).

Department of Trade and Industry (DTI): UK Government department with the overall aim   'To increase competitiveness and scientific excellence in order to generate higher levels of sustainable growth and productivity in a modern economy '. (Website: www.dti.gov.uk).

depth cueing: A *computer graphics* technique which gives an appearance of depth in a display by varying the intensity levels of the displayed objects.

depth of field: The range of object distances from a camera within which objects will be reproduced as acceptably sharp images. See   **Figure D.3**.

depth of focus: In an optical system, the range of distances in front of and behind the *image plane* within which the image is acceptably sharp. See *circle of confusion* and **Figure D.3**.

deserializer: An electronic device which converts a serial data stream to a parallel one (compare *serializer*).

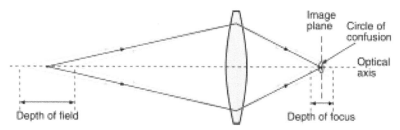

Figure D.3   Depth of field and depth of focus

desk-top editing: Video *editing* and production carried out on a standard desktop computer with add-on video hardware and running one or more *software* applications. Also called *desk-top video.*

desk-top publishing (DTP): The use of a *computer workstation* to develop page layouts for subsequent printing.

desk-top video: See *desk-top editing.*

detail: The smallest *elements (5)* in a television picture which are distinct and recognizable. Similar to *definition* or *resolution.*

deuce: A 2 kW *Fresnel lens* spotlight.

Deutsches Institut für Normung (DIN): (1) German national standards organization (Germany). **(2)** The term used to denote photographic film sensitivity. See also *speed.*

developer: The chemical used for *developing* an exposed photographic film.

developing: The process by which the *latent image* in an exposed photographic film is made permanently visible.

deviation: The range of frequencies over which the *carrier* frequency is swept or *modulated* in a system using *frequency modulation.*

dia: See *diapositive.*

diagnostics: Programs built into a device which test its functionality and report the results as an aid to fault finding.

dialect: A variation of a 'standard' *computer language.*

diagonal wipe: A *wipe* in the form of a triangle from one corner . See **Figure W.1.**

diamond wipe: A *wipe* in the form of a diamond. See **Figure W.1.**

diaphragm: (1) The opening in an optical system controlling the amount of light transmitted. In an *iris* diaphragm the transmission is adjustable. **(2)** A membrane used in certain *microphones* and *loudspeakers.*

diapositive: French and German word for a *transparency.*

diascope: A still *transparency projector.* (Hist.)

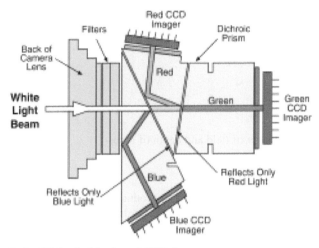

Figure D.4   Dichroic block and CCD imagers

diazo: An imaging process based on diazonium compounds used in *overhead projector* transparencies, *slides*, prints and the preparation of photo-resist materials.

dichroic: Having selective reflection and transmission for *electromagnetic waves* of certain *wavelengths*.

dichroic block: The colour beam-splitting system of a professional three-chip video camera on which the *CCD imagers* are fixed. See **Figure D.4**.

dichroic filter: A *filter (2)* or mirror which transmits a certain spectral range and reflects others. Examples are the video camera *dichroic block* or the separation of visible light and *infra-red* (heat) in a projection system.

dichroic head: A colour light source based on adjustable *dichroic filters*, used with photographic *enlargers* and *rostrum cameras*.

Didon: A French *teletext* service.

dielectric: An electrical insulator.

difference frequency: A third frequency produced by the interaction of two different fundamental frequencies.

differential amplifier: A device which amplifies the dif ference between two input signals.

differential gain: Often abbreviated to 'diff gain', it is a measure of variations in *saturation* which may occur with increasing *luminance* level. It is measured using a *staircase test signal*, but can be estimated using a *vectorscope* display of *colour bars*.

differential input: A system with two inputs of equal sensitivity which are combined in *anti-phase*. This results in an output proportional to the difference between the inputs.

differential phase: Often abbreviated to 'diff phase', it is a measure of variations in *hue* which may occur with increasing *luminance* level. It is measured using a *staircase test signal*, but can be estimated using a *vectorscope* display of *colour bars*.

differential pulse code modulation (DPCM): *Pulse code modulation* in which the differences between samples is coded rather than the actual sample values.

diffraction: (1) A change in the direction of a wave at the edge of an obstacle in its path. **(2)** *Interference* patterns produced when a wave (water, sound, light or radio) passes around an opaque obstruction or through an opening of roughly the same dimensions as its *wavelength*.

diffuser: A translucent *filter (1)* placed in front of a light source to soften the resultant shadows, or placed in front of a camera lens to reduce the sharpness of the image.

diffusion: (1) The scattering of light. **(2)** The distribution of reflected sound waves in an enclosed space.

Digicipher: A proprietary version of *MPEG-2* used by General Instruments for *satellite (2)* programme distribution. (Trade name.)

digit: A single number.

digital: Expressed in numerical terms, often in *coded (1)* form, as opposed to *analogue*.

Digital 8: A consumer digital videotape recording format from Sony using an 8mm Hi-8 type tape cassette. Sometimes also called *D-8*.

digital audio broadcasting (DAB): A digital audio transmission system using *COFDM* techniques to *broadcast* several digital radio services in a single transmission *channel (2)*.

digital audio stationary head (DASH): A digital audio tape recording system.

digital audio tape (DAT): A two-channel digital audio recording system in two version: *R-DAT* with rotating *heads (2)*, *S-DAT* with stationary *heads (2)*. DAT is usually taken to mean *R-DAT*.

Digital Audio Visual Council (DAVIC): A worldwide consortium based in Geneva whose members represent all industries involved in the audio and video applications of digital technologies. One of its aims is to define open *interfaces (1, 2)* for maximum uniformity across countries, applications and services. (Website: www.davic.org).

Digital Betacam: A *digital videotape recorder* format from Sony which uses video data compression of about 2 :1 using a proprietary data compression system based on *JPEG* techniques.

digital component: *Component video* where the components are in their *digital signal* form.

Digital Component Technology (DCT): *DVTR* format developed by Ampex which records a *compressed* digital *component video* signal on a 19mm ($\frac{3}{4}$ inch) *D2* type tape cassette.

digital delay: A signal delay system which works by storing numbers representing the signal.

digital disk r ecorder (DDR): A *hard disk drive* system which records digital video, often used as a replacement for a *videotape recorder* or as a video *cache (2)*.

Digital Interface (DIF): The data structure shared by all *DV* based compression schemes and defined in the *Blue Book (1)*.

Digital International Conversion Equipment (DICE): The first digital system for *broadcast* television *standards conversion* developed by the *IBA*. (Hist.)

Digitally Assisted Television (DATV): European proposal for reducing the bandwidth of *high definition television* for transmission by a single *DBS* channel. (Hist.)

digital multi-meter (DMM): A *multimeter* with a digital display of the measured values rather than a *moving-coil* meter.

digital panel meter (DPM): Panel mounting meter with a digital display of the measured values rather than a *moving-coil* meter.

digital pitch changer: A device using *digital* techniques to alter the *pitch (2)* of the original sound. It is used with a small *pitch (2)* change to reduce the onset of *threshold howl*.

Digital Production Effects (DPE): An early *DVE* made by *Quantel* which could carry out a range of simple two-dimensional effects. (Trade name.) (Hist.)

Digital-S: See *D9* (Matsushita trade name).

Digital Satellite System (DSS®): Proprietary digital satellite broadcasting system used by some US operators, based on *MPEG-2*, but incompatible with DVB-S. See **Appendix S**.

digital signal: A signal with discrete values. These are usually *coded (1)* to carry information representing, for example, audio or video (compare *analogue signal*).

digital signal pr ocessing (DSP): Processing a signal using digital integrated circuits.

digital tape r ecording system (DTRS): A multi-track audio system recording up to 8 digital audio tracks on an 8mm *videotape cassette*.

Digital Television Group (DTG): A group of interested parties committed to launching digital television services in the UK. (W ebsite: www.dtg.co.uk).

digital terrestrial television (DTT): Using complex digital transmission techniques to *broadcast* several different television services to *viewers (2)* in the same *ultra high frequency* channel.

Digital Terrestrial Television Action Group (DIGITAG): An organization aiming to ensure the 'harmonious introduction of European *DTT* services to the mass consumer'. (Website: www.digitag.org).

Digital Theater Systems (DTS): A multi-channel film sound system using *ADPCM* predictive audio coding. 5.1 channels (left, centre, right, left-surround, right-surround and *sub-woofer*) are encoded at 4:1 data compression with 16-bit *quantization* and 44.1 kHz *sampling*. The data is recorded on *CD-ROM* and synchronized with the replayed pictures using an optical time track recorded either between the optical sound track and the image for 35mm film prints or outside the perforations on 70 mm film prints (see **Appendix A**). The system was first used for the 1993 film 'Jurassic Park'. (Website: www.dtstech.com).

Digital Theater Systems – Extended Surround (DTE-ES): *DTS* with an additional centre surround speaker.

digital-to-analogue converter (DAC): A device that converts a coded *digital signal* into its equivalent *analogue signal*.

Digital Versatile Disk (DVD): An optical *disc* of 12cm diameter used to record digital audio, video or data and read by an optical *laser* system. It can be single or dual layered and/or single or double-sided and has a far greater data storage capacity than a *compact disc*. A single-sided, single layer *disc* has a capacity of 4.38 gigabytes. It was originally called *Digital Video Disc*. See **Appendix Y** for more details. (W ebsite: www.dvda.org).

Digital Vertical Interval Time Code (DVITC): *Time code* signal carried in the *vertical interval* of a digital video signal and defined in standard SMPTE 266M. See **Appendix U**.

Digital Video Broadcasting (DVB): A group with over 200 members in 25 countries. They developed and gave their name to the standards for all digital television *broadcasts* in Europe. See **Appendix S**. (Website: www.dvb.org).

Digital Video Disc: See *Digital Versatile Disc.*

Digital Video Effects (DVE): Registered trademark of Nippon Electric Company, now used generally to refer to video equipment which carries out complex television picture manipulations such as squeezes, *zooms, tumbles* and spins.

Digital Video Interactive (DVI): A standard for integrating digital video, graphics and sound in a compressed form, used in interactive multimedia applications. It was originally developed by RCA, then sold to Intel in 1988.

Digital Videotape Recorder (DVTR): A *videotape recorder* which records/replays *digital* signals. The first *D1* format DVTR for commercial use was shown in 1986.

digitize: (1) Converting an image from hard copy into digital data for use in a computer. **(2)** Converting an analogue signal into digital values. **(3)** Capturing video material onto the *hard disk drive* of a *desktop editing workstation* prior to beginning the *editing* process.

digitizer: A device to aid the input of digital information on a *computer*, e.g. dimensions from a drawing. Examples are mice, *data tablets* or *graphic tablets*.

dimmer: An electrical or electronic device for controlling lamp intensity.

DIN: (1) A number indicating photographic film sensitivity. See also *ASA (2)*, *exposure index* and *speed*. **(2)** German standards body, *Deutsches Institut für Normung*.

DIN plug: A range of multi-pin connectors standardized by *DIN (2)*. These circular connectors and the 2-pin *loudspeaker* connector are frequently found on *audio-visual* equipment.

diode: An electronic component which allows transmission of electric current in one direction only.

dioptre: (1) Unit of magnifying power of a lens, the reciprocal of its *focal length* in metres. **(2)** A single *supplementary lens* used in front of the main camera lens for close-up photography.

dip: A metal trap in a floor for electrical sockets.

dip filter: An audio *parametric equalizer* with a very narrow *bandwidth*, used to greatly reduce the effect of camera noise, etc. which has a narrow frequency range.

dipole: An *aerial* having two *elements (2)* of equal length, fed from the centre.

dip stick: The mounting for a *gobo (1)* on a *luminaire*. (Colloq.)

direct current (DC): Electric current which keeps a constant polarity (see *alternating current*).

direct cut: A vinyl *audio disc* cut 'live' without the intermediate use of magnetic recording. (Hist.)

direct broadcast by satellite (DBS): Transmitting many television services to *viewers (2)* over a large geographic area by using a single *satellite (2)* in *geostationary Earth orbit* and a small exterior receiving *aerial* (dish) and *set-top box* at each *viewer's (2)* home.

direct exchange line (DEL): A dedicated telephone line between an *OB truck* or a studio *control room* and a telephone exchange. (Hist.)

directional microphone: A *microphone* which is more sensitive in certain directions. Its *polar diagram* will generally take one of three forms: a figure of eight for a *bidirectional microphone*, a heart shape for a

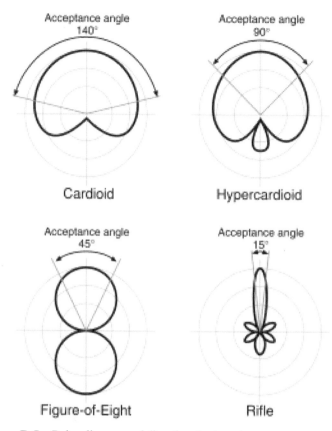

Figure D.5 Polar diagrams of directional microphones

cardioid microphone or between the two for a *hypercardioid microphone*. Also called a *unidirectional microphone*. (Compare *omnidirectional microphone*.) See **Figure D.5**.

direct memory access (DMA): A method of fast direct input or output of data to or from a *computer* memory, bypassing the *central processing unit*.

director: A short horizontal, vertical or X-shaped bar on a *Yagi aerial* used to improve its directivity.

Directors Guild of America (DGA): Professional society for US film and video directors. (Website: www.dga.org).

direct read after write (DRAW): A replay technique used in a data system such as a recordable *videodisc*.

71

direct-to-home (DTH): Reception of *satellite (2)* television *broadcasts* directly in a subscriber's home.

disc: Describes optical storage media, e.g. *compact disc* or *videodisc*. (Compare *disk* for magnetic storage media.)

discrete cosine transform (DCT): A mathematical process. When applied to digital video compression it is used as a reversible technique by which the information contained in a small area of a digital picture (often 8 × 8 *pixels*) is translated from one form (sample values) to another (frequency coefficients) which are more suited to subsequent *bit rate reduction* processing.

dish: (1) A concave *aerial* for *microwave* transmission or reception. See **Figure P.1**. (Colloq.) **(2)** A reflector used with a form of directional *microphone*.

disk: Describes magnetic storage media, e.g. *hard disk drive* or a *floppy disk* (compare *disc* for optical storage media).

disk crash: A *computer* failure resulting from a *hard disk drive* read error. If the head *(2)* actually touches the magnetic surface, both the *head (2)* and the *disk* can suffer physical damage and will therefore almost certainly be irretrievably damaged.

diskette: See *floppy disk*.

disk file: *Computer* data stored on a magnetic *disk* under a *file name*.

disk operating system (DOS): A *computer program* which includes routines for managing *hard disk drive* or *floppy disk* access.

dispersion: (1) The separation of light into its constituent *wavelengths*, as in white light passing through a prism. **(2)** The process of mixing the magnetic particles and other constituents of the coating during the manufacture of *magnetic tape*. **(3)** The variation of transmission time in a medium, particularly in *optical fibres* the spreading out in time of an optical pulse.

dissolve: A visual *transition* between two pictures in which the whole area of the first gradually disappears as it is replaced by the second. Also called a *mix*.

dissolve cycle time: In *slide* projection, the *dissolve time* plus the time taken for the next *slide* to be positioned ready for projection.

dissolve pair: Two *slide projectors* arranged to allow *dissolves* between *slides*.

dissolve time: The period taken to complete a *dissolve*.

distortion: (1) In electronics, the generation of unwanted signals such as *harmonics* or *intermodulation* products. **(2)** A reduction in the quality of a reproduction as compared to the original.

distribution amplifier (DA): An *analogue* circuit for amplifying an audio or video signal. It often provides several separate outputs and can be used as an *equalizer* for a signal path.

dither: **(1)** The intentional addition of low level *noise* in a *digital videotape recorder* system, to reduce the visibility of *quantizing* effects. **(2)** In a mechanical system, an intentional introduction of a small movement to overcome sticking. **(3)** In *computer graphics*, an *anti-aliasing* technique in which colour or intensity variations are introduced.

D-multiplexed analogue component (D-MAC): An analogue component television transmission system. There are two varieties: D-MAC and D2-MAC (not to be confused with the *D-2* digital *VTR* standards). See *multiplexed analogue component*. (Hist.)

dock: **(1)** An area used for temporarily storing scenery . **(2)** To join a *videotape recorder* back to the body of a video camera to form a one-piece unit.

Dolby®: Dolby Laboratories, named after their founder Ray Dolby , have invented and given their name to a variety of *noise reduction* and *surround sound* systems. (Website: www.dolby.com).

Dolby A, B and C: These *noise reduction* systems are found in professional and domestic *analogue* audio and *videotape recorders*. They process the sound signal before recording and during replay to give less audible *noise* than would otherwise be the case. Dolby A was introduced in 1965 for professional sound recording and splits the signal into four frequency bands for processing (trade name). Dolby B was introduced in 1968 for consumer systems and only reduces tape hiss above 5 kHz.

Dolby Digital: Originally called *AC-3*. A multi-channel digital sound system working with between 1 and 5.1 channels. It is a *constant bit rate* system using between 64 kbit/s and 640 kbit/s. The 5.1 channel 35 mm film version records the digital information optically between the *perforations*. (See **Appendix G**.) It was first used for the 1992 film 'Batman Returns'.

Dolby E: A way of digitally encoding a 5.1-channel mix for digital television and a *Dolby Surround* mix for analogue television in a single two-channel AES/EBU digital audio signal. Audio *metadata* is also carried to optimize consumer playback.

Dolby level: A reference signal *amplitude* used to adjust the *Dolby noise reduction* system for correct decoding.

Dolby SR: *Dolby* Spectral Recording. A sound recording process that combines the attributes of the *Dolby A and C* type *noise reduction* systems with other processing. It is used in the highest quality professional analogue audio and *videotape recorders*. (Trade name.)

Dolby SR-D: A film release format incorporating a *Dolby SR* analogue track and a 6-channel digital track.

Dolby Stereo: A *surround sound* system for *encoding* and recording four channels of sound information (left, right, centre front, and surround) onto two analogue tracks of a film recording. These can be *decoded* and sent to multiple speakers in a cinema *auditorium* for dramatic sound effects. The first Dolby Stereo film was the 1975 film 'Lisztomania'.

Dolby Surround: The domestic version of *Dolby Stereo,* which allows sound tracks recorded in *Dolby Stereo* to be heard in *surround sound,* using fewer *loudspeakers* and simpler signal processing. Dolby Surround Pro Logic uses more complex signal processing to give an improved effect and has a separately derived centre speaker output.

Dolby Surround EX: A digital sound release format developed by Dolby Laboratories and THX for the 1999 film 'Star Wars: Episode One – The Phantom Menace'. It has three surround speaker channels derived by matrix encoding them in the two previous surround tracks.

dolly: A movable platform onto which a camera may be mounted so that *action (2)* may be followed, hence dollying. See also *tracking (1).*

Dolphin arm: A short counterbalanced camera *boom* arm made by Vinten, which can be mounted on various types of *dolly.* (Trade name.)

domain name: The name of a group of computing and storage systems on a distributed *network*, usually the *Internet*, and identified by groups of characters separated by full-stops. See also *Internet domain.*

domain name service (DNS): *Internet* service that translates *domain names* into *IP* addresses. This makes access for the *end-user* easier since *domain names* tend to be based on acronyms, whilst *IP* addresses are numerical. See also *Internet domain.*

dope sheet: A breakdown of instructions for shooting scenes. See also *story board.*

Doppler effect: The perceived change in *wavelength* where there is relative motion between a wave-energy source and an observer. Sound, for example, suffers a change in pitch, light a change in colour and *radio frequency* radiation a change of *frequency.*

dot: A very small circular *flag (1)* to control part of the light beam from a *luminaire.*

dot crawl: The pattern of moving dots seen on a video picture around the edges of objects with highly saturated colours. It occurs with *composite video (2)* and is due to the *decoder* mistaking colour information for *luminance.* Hence it is also called *cross luminance* (compare *cross colour*).

dot matrix: Forming *alphanumeric* characters within a standard pattern of dots.

dot matrix printer: A low-quality printer using an electromagnetically operated array of small pins (usually between 9 and 24 arranged in a line) which marks the paper by pressing the pins on an inked ribbon. Characters or graphics are printed as a pattern of dots as the print head moves across the paper.

dot pitch: The distance between dots or stripes of the same colour in a colour display, measured and quoted in millimetres.

double density: A system which records (usually on *floppy disks*) with a higher density to store more data on a given area of medium.

double duplex sound track: A photographic sound track having two *duplex* sound tracks side-by-side *in phase*, within the standard track width.

double-edge sound track: See *double-sided sound track*.

double exposure: *Exposure* of the same piece of photographic film twice to record two dif ferent images. These may be superimposed or may occupy areas reserved by *mattes*.

double frame: A system of motion picture photography in which the image area is twice that generally taken as standard.

double frame animation: Each drawing or diagram in a sequence is photographed twice, usually to save labour .

double-headed: Film projection or similar equipment capable of accepting separate picture and sound films, and running them in *sync*.

double-sided sound track: A photographic sound record with a dif ferent sound track on each edge, recorded in opposite directions. Also called a *double-edge sound track*.

double system: In *cinematography*, the production process in which picture and sound are recorded on separate films or on film and tape.

double tracking: A second identical performance of a previously recorded video line.

downconverter: Electronic equipment used to reduce the number of television lines from *high definition television* to *standard definition television* (compare *upconverter*).

downlink: In a *telecommunications satellite (2)* system, the return path from the *satellite (2)* to the receiving *dish (1) aerial* (compare *uplink*).

down rate: In *slide* projection, the time taken from the start of a *fade* to the disappearance of the light of the down-going *projector*.

downstage: Performing area nearest the camera or audience.

downstream keyer: A *caption* superimposer on the output of a vision mixer. It allows *titles* to be added to the mixer output picture and is not affected by the preceding mixer controls.

downtime: The length of time a system or unit is inoperative due to maintenance or repair.

dowser: A *shutter (3)* used to cut off the light beam in a *projector* or film printer.

dragging: Moving a *computer graphic* object along a path determined by a graphic input device.

drapes: Curtains.

dressing light: See *background light*.

drift: The gradual variation of a parameter, e.g. *oscillator frequency* or tape speed.

drop-frame time code: US *time code* system which compensates for the difference between the NTSC *frame rate* of 29.97 per second and *time code* counting 30 *frames* for each second. Two *frame* numbers every minute are dropped except during the tenth minute.

drop-in: In audio recording, the inserting of a replacement section into an already recorded piece by switching the required channel to record at the correct position along the tape. Similar to *insert* in video *editing*.

dropout: A short loss or severe deterioration in the replayed signal from a *videotape recorder* usually due to loss of *head*-to-tape contact or physical faults in the *magnetic tape*.

Dropout Compensator (DOC): An electronic circuit that reduces the subjective effect of a *dropout* by replacing the part of the replayed picture affected with video from part or the whole of a previously stored line of the picture.

drop shadow: A dark edge on lettering and other objects giving the appearance of relief. The effect may be produced photographically or electronically as well as in original artwork.

drum: (1) A rotating cylinder round which film passes to control its uniformity of movement in a photographic sound reproducer . **(2)** See *head drum*.

dry hire: Hire of equipment without operating personnel.

dual cone: A type of *loudspeaker* with two concentric cones driven from one *voice coil*.

dualateral sound track: A photographic sound track with two in-phase unilateral variable area records with *modulations* in the same direction.

dual-in-line (DIL): An *integrated circuit* package with the contact pins in two parallel rows for mounting through holes in a circuit *board (1)*.

dubbing: (1) Combining two or more sound records into a single composite recording. **(2)** Transferring a sound recording from one medium to another, e.g. from *magnetic tape* to photographic film. **(3)** Making a copy, e.g. a show copy cassette from a reel-to-reel tape. **(4)** In motion picture production, the process of recording new dialogue to be substituted for the original version.

dubbing switch: A switch on *video cassette recorders* used to improve the quality of tapes being copied.

dummy load: A device which absorbs all the power of a transmitter or amplifier for testing or standby operations.

dumping: Saving data from electronic memory on to a permanent medium, for example to *floppy disk* or data *cassette* tape.

duo-bilateral: The variable-area photographic soundtrack format used for almost all 35 mm mono and stereo film soundtracks.

dupe: Abbreviation for *duplicate*.

duplex: **(1)** A mode of data transmission where each station can send and receive data simultaneously. **(2)** Simultaneous *videotape* recordings of the same source with identical *time codes*.

duplex sound track: A photographic sound track with two in-phase unilateral variable area records having *modulations* in opposite directions.

duplicate: **(1)** For *slides*, audio tapes and *videotapes*, a reproduction matching the original as closely as possible. **(2)** In motion pictures, a *copy negative* matching the original exposed in the camera.

duplication: **(1)** In photographic practice, the preparation of a film or *slide* duplicate from original material. **(2)** In audio and videotape usage, the preparation of copies in general.

dutch dolly: A three-wheeled steerable *camera mount* with rubber tyres often used on *outside broadcasts*.

DV: Digital video. The name of a digital videocassette format developed for the consumer market. It records a *compressed* digital *component video* signal at 25 Mbit/s on a 6.75mm ($\frac{1}{4}$ inch) tape cassette using *metal evaporated* tape. Upgraded versions including *DVCPRO* and *DVCAM* are aimed at the professional and broadcast markets. The name is also applied to the video data compression system used in these formats. Defined in standard SMPTE 314M (see **Appendix U**).

DVB-C: DVB framing, *channel coding* and *modulation* scheme for digital *cable television* systems. See **Appendix S**.

DVB-S: DVB framing, *channel coding* and *modulation* scheme for digital *satellite (2)* television systems. See **Appendix S**.

DVB-T: DVB framing, *channel coding* and *modulation* scheme for *digital terrestrial television* systems. See **Appendix S**.

DVCAM: *DVTR* system developed by Sony which records a *DV compressed* digital *component video* signal at 25 Mbit/s on a 6.75mm ($\frac{1}{4}$ inch) tape cassette using *metal evaporated* tape.

DVCPRO: See *D7*. (Matsushita trade name.)

DVCPRO50: *DVTR* format, developed by Panasonic from *DVCPRO*, which records a *DV compressed* digital *component video* signal at 50 Mbit/s on a 6.75mm ($\frac{1}{4}$ inch) tape cassette using *metal particle* tape.

DVD Association (DVDA): A non-profit making organization open to all those connected in any way with the creation of *Digital Versatile Disc* titles. (Website: www.dvda.org).

DVD Forum: An international association of hardware manufacturers, software firms and other users of *Digital Versatile Discs*, created to exchange and disseminate ideas and information about DVD

Figure D.6   Dynamic tracking head

technical   capabilities, improvements and innovations. (W   ebsite: www.dvdforum.org).

D-weighting: A frequency *weighting* network sometimes used for the measurement of aircraft *noise*.

dynamic microphone: A *moving-coil* type of *microphone*.

dynamic random access memory (DRAM):   *Random access memory* where the information is lost unless constantly refreshed, by either recirculating the data or in some devices just by addressing it (compare *static random access memory*). It was invented in 1966 by an American, Robert Dennard.

dynamic range: The usable range between the strongest signal capable of being handled by a device and the inherent   *noise*.

dynamic track following (DTF): Used in *high-band videotape* replay to obtain perfect *tracking (4)* in slow-motion or *still frame*.

dynamic tracking (DT):   Used in *videotape* replay to obtain perfect *tracking (4)* in reverse, slow-motion or   *still frame* (Sony). See also *AST*.

dynamic tracking head: A *videotape* replay *head (2)* which provides a *dynamic tracking* facility, usually using a *bimorph*. See **Figure D.6** and **Figure B.3**.

dyne: Unit of force. The force needed to accelerate a mass of 1 gram at 1 cm s^{-2}. The preferred *SI unit* is the *newton*, where 1 newton = 10 000 dynes.

E

earpiece: The part of a telephone handset which is put near to the ear .

earth: A conductor physically connected to the ground. (US usage is simply *ground*.)

earth leakage cir cuit breaker (ELCB): Obsolete term for a *residual current device*. (Hist.)

earth loop: A fault condition affecting signals passed between units where circulating currents in multiple ground paths introduce *hum*.

earth station: A transmitting and/or receiving station on Earth forming part of a *satellite (2)* communications system. Also called a *ground station*.

earwig: A small *earpiece* used to give actors a guide track, often music. (Colloq.)

echo: (1) The return of characters from a keyboard to a monitor *screen (1)*: local echo coming direct from the point of origination, remote echo from a distant system. **(2)** A discrete separately identifiable repetition of a sound after reflection from a hard surface (compare *reverberation*).

echo unit: In audio recording, equipment for generating time-delayed signals to be added to the original sound.

edgecodes: See *edge numbers*.

edge numbers: Groups of sequential numbers, sometimes including letters, along the edge of motion picture film at regular intervals, usually one foot apart. Also called *footage numbers* or *edgecodes*. Defined in standard SMPTE 83 (see **Appendix U**).

edgewave: A film distortion in which the edge of the film is stretched compared with the centre. Also called *fluting*.

Ediflex: A *VHS* band non-linear video *editing* system. (Trade name.)

edit decision list (EDL): A list of the source and *time code* of *videotape* edits. It may also contain supplementary information.

editing: The process of selecting from a large amount of recorded material and assembling a programme by combining separate sequences.

editing block: A slotted holder for *magnetic tape* to enable physical *joins* to be made more accurately .

edit point: In film or video *editing*, the precise *frame (1, 2)* at which an edit is chosen to take place.

Educational Television & Media Association (ETMA): UK organization bringing together users of video and *screen*-based media in education and training. (Website: www.etma.org.uk).

Effective Isotropic Radiated Power (EIRP): The total power supplied to an *aerial* multiplied by the gain of the *aerial* in a given direction.

effects (FX): Can refer to sound ef fects, *special effects (1)* or optical effects.

effects track: A sound track containing sound ef fects only.

egg boxing: The division of a multi-image *screen (1)* into compartments so as to provide a sharp edge between each projected section. Also called *egg crating.*

egg crate: A lattice of horizontal and vertical slats fitted to the front of a *luminaire* to restrict the spread of light.

egg crating: See *egg boxing.*

Eidophor: A powerful video projection system using light from a separate source passing through a film of oil that is distorted by a *scanning* electron beam. (Trade name.)

eigentones: Preferential frequencies at which acoustic *resonances* occur, related to the separation of parallel walls in a room.

electret microphone: A *microphone* working in the same way as a *capacitor microphone* but with a permanent char ge on the foil so that it does not need a polarizing voltage supply .

Electrical, Electronic, Telecommunications and Plumbing Union (EETPU): UK trade union which mer ged with the *AEU* to form the *AEEU*. (Hist.)

Electromagnetic Compatibility (EMC): European Union directive covering the generation and ef fects of *electromagnetic interference.*

electromagnetic deflection: Deflecting an electron beam in a *cathode ray tube* using an external magnetic coil and used when lar ge deflection angles are needed, such as in a video picture monitor (compare *electrostatic deflection*).

electromagnetic interference (EMI): Unwanted *electromagnetic waves* which radiate unintentionally from an electronic circuit or device into other circuits or devices which may disrupt their operation.

electromagnetic spectrum: The total range of electromagnetic radiation including radio waves, *microwaves, light, X-rays* and *gamma rays*. See **Appendix N.**

electromagnetic wave: A wave produced by the interaction of an electric wave and a magnetic wave. Examples are light and radio waves. See also *polarization* and **Figure E.1.**

electromotive force (EMF): A difference in electrical potential which gives rise to a flow of current, measured in *volts.*

electron beam recorder (EBR): A high quality *videotape* to film transfer system. (Hist.)

electron gun: The device in the neck of a *cathode ray tube* which provides a defined beam of electrons.

electronic balanced input: A system where the *balanced* circuit is obtained using *active components* rather than centre-tapped *transformers.*

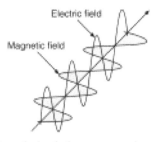

Figure E.1   Vertically polarized electromagnetic wave

electronic editing: The selection and composition of video and/or audio information, making a new recording by transfers from a variety of original material.

electronic field production (EFP): Location recording, usually for drama, using portable television cameras and *videotape recorders*, to *broadcast standard.*

Electronic Industries Association (EIA): Trade organization based in Washington, DC, which represents US electronic equipment manufacturing companies. They helped define the 525-line television standard. See also *CEMA, JEDEC* and *TIA*. (Website: www.eia.org).

Electronic Industries Association of Japan (EIAJ): Founded in 1948 as a non-profit making national trade or ganization to develop Japan' s electronics industry and represent its views. (W ebsite: www.eiaj.or.jp/ english/index.htm).

electronic news gathering (ENG): A technique for capturing news events in the field using lightweight, portable battery-powered television cameras and *videotape recorders.*

electronic pin registration (EPR): A system that ensures consistent *frame (1)* steadiness in *telecine* transfers.

electronic post-production (EPP): See *post production.*

electronic programme guide (EPG): In digital television, a database held in a viewer 's digital receiver or *set top box* , which is updated from the broadcaster 's own database by transmitting special data *packets* with the broadcasts. This allows a receiver to build an on-screen display of continuously updated programme information and contains control information to facilitate navigation and selection. An EPG therefore comprises a database and an on-screen navigable display.

electronic publishing: Images reproduced through television, video devices, *optical discs*, tapes, cassettes, *closed circuit television broadcasts*, cable *broadcasts* and other electronic means.

electronic-to-electronic (E-to-E): Used to describe a *videotape recorder* operating mode where the input signal is processed by all the electronic systems but is not recorded.

electrostatic: See *static (2)*.

electrostatic deflection: Deflecting an electron beam in a *cathode ray tube* using a high voltage between a pair of internal metal plates and used when small deflection angles are needed, such as in an *oscilloscope* or *waveform monitor* (compare *electromagnetic deflection*).

element: (1) The part of a *microphone* which converts sound energy into an electrical signal. **(2)** Part of an *aerial* system. **(3)** An individual lens component in a complex photographic lens. **(4)** A small functional component in a complex circuit, such as a *CCD imager* or other *integrated circuit*. **(5)** The smallest part of an image, usually video, which can be separately observed or processed, as in *pixel*. **(6)** In *MPEG* video, the part of a programme representing video, audio or data.

elementary stream (ES): *MPEG* term for the output data stream from a video or audio data *compression (2) encoder*.

elevation: (1) The height of a location above mean sea level. **(2)** The number of degrees by which an *aerial* is tilted upwards from the horizontal plane.

elevator: (1) An arrangement of fixed and moveable film rollers acting as a variable capacity reservoir for a film processing machine. **(2)** The robotic mechanism for transporting *cassettes* in a video *cart machine*.

elvis: In lighting, a frame covered with gold lamé and used to bounce light.

e-mail server: A *server* dedicated to providing internal (and often external) electronic mail and messaging services on a *network*. Also called a *mail server*.

embedded audio: Digital audio signal(s) which are *multiplexed* onto a serial digital video data signal. *SDI*, for example, carries four channels of embedded digital audio.

emission: The means by which radio waves are propagated into the atmosphere or space.

emulsion: In photographic materials, the layer containing the light-sensitive agents in a medium such as gelatine.

encoding: (1) Putting information into a prescribed coded form, suitable for transmission. **(2)** In *slide* projection, adding *cue tones* or other control signals to an already prepared audio tape.

encoder: A device which converts a signal from one form into another. For example, a shaft encoder converts angular position or revolutions into digital form. In television, a particular application is a device used to form a single *composite video (2)* signal from a set of *component video* signals. Another application is to reduce the data rate requirements for digital video in a data *compression (2)* encoder, e.g. in *MPEG-2*.

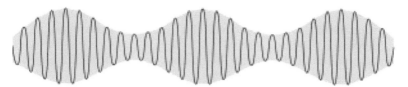

Figure E.2   Envelope of a wave

encryption: *Scrambling* a television transmission so that it can only be reproduced at stations or receivers equipped with a corresponding decoder and *smart-card*, as applied in *pay TV.*

end board: The use of an inverted *clapper board* at the end of a scene.

endless cassette: An audio *cassette* containing an endless loop of tape, which can run only one way and cannot be rewound.

end user: The actual user of a system.

engineering cue: An additional cue provided with certain *tape-slide* systems to advance the *slide projector magazines (2)* to the first programme *slide.*

engineering pulse: A recorded command, cue tone or signal which checks the operation of a system, sets the parameters of operation or advances the equipment to the start position.

enhancement: In television, increasing the apparent *resolution,* either horizontally or vertically. See *aperture correction.*

enlarger: Photographic equipment used to expose photo-sensitive paper to an image from a positive or negative film.

entitlement management message (EMM): Used in *MPEG-2* to carry *keys* to allow *encrypted* programmes to be descrambled and viewed by permitted users.

envelope: (1) The enclosure containing the components of a lamp, *cathode ray tube* or similar vacuum device. **(2)** The boundary of the highest and lowest *amplitude* excursions of a signal, such as a variable-area film sound record or an of f-tape *RF* signal. See **Figure E.2.**

epidiascope: An optical *projector* which combines the function of a *transparency projector (diascope)* with that of an *episcope.* (Hist.)

episcope: An optical *projector* producing an image of opaque objects, such as photographs or pages of a book, by reflection from their surfaces. (Hist.)

equalization: (1) In audio, the use of frequency-selective *attenuators* and/ or *amplifier* circuits to change the relative balance in *amplitude* between higher and lower frequency ranges. **(2)** Altering the frequency response of a video amplifier to compensate for high-frequency losses in *coaxial* *cable*. For serial digital video, equalization is often carried out automatically by the input circuit of the equipment receiving the signal.

equalizer: A device for altering the *frequency response* (and perhaps phase response) characteristic of a signal.

equalizing pulses: A group of five *pulses* at twice line rate inserted either side of the vertical sync pulse period in the *vertical interval*. They were originally included to maintain scanning stability in television receivers by ensuring that both fields had the same format in the region of the field sync pulse period. See **Appendix L**.

equipment rack: A metal frame structure or cabinet with mounting strips to secure electronic equipment at fixed spacings based on rack-height units (*U*).

erase: To remove a previous recording from *magnetic tape* or film.

error correction: Putting right errors which have occurred and been detected. At a *bit* level this just means inverting the sense of the *bits* in error.

error correction codes (ECC): The extra codes that need to be added to a data stream to enable *error correction* to be carried out.

error detection and handling (EDH): The error strategy used by a data system. It involves the reliable detection of errors and how to deal with them, either by *error correction* or masking them by *interpolation* if the data is audio or video originated.

errored second: A second of time in which one or more errors are detected. It is used as a measure of the performance of a digital system, e.g. digital video on an *SDI* connection (compare *bit error rate*).

estar: Polyethylene terephthalate base. See *mylar*. (Trade name.)

Ethernet: A *local area network* invented by Bob Metcalfe in 1972 and developed at Xerox in 1976 for linking *personal computers*. It was adopted in 1980 by DEC, Intel and Xerox as a standard communications system for *personal computer* systems. The name originates from the first system linking Washington, Palo Alto and Seattle with radio links (so using the ether!). Ethernet is a trademark of Xerox Corporation, Inc.

Euroconnector: See *SCART connector*.

European Broadcasting Union (EBU): Part of the *World Broadcasting Union*, based in Geneva, Switzerland. It is a private non-profit association of 66 radio and television *broadcast* organizations in Europe and the Mediterranean area, with 51 associate members in other parts of the world. It negotiates broadcasting rights for major sports events, organizes programme exchanges, operates the *Eurovision* and *Euroradio* networks and stimulates and co-ordinates co-productions. (W ebsite: www.ebu.ch).

European Telecommunications Satellite Organization (EUTELSA T):
A *satellite (2)* operator based in France with 15 *satellites (2)*
broadcasting more than 550 digital and analogue television channels to
over 75 million satellite and cable homes in Europe, Africa and Asia.
(Website: www.ebu.ch).

European Telecommunication Standard (ETS): A standard published by
ETSI.

European Telecommunications Standards Institute (ETSI): EU organization (European Union) based in France, set up to create and harmonize
telecommunications standards across Europe. It has 490 members from
34 countries. (Website: www.etsi.org/eds).

European Teleconferencing Federation (ETF): Promotes the teleconferencing industry including audio and video conferencing, datashar-
ing and business television using ISDN, Internet and satellite
technologies.

Euroradio: An arrangement made between European and Mediterranean
television services for the distribution of radio programmes.

Eurovision: An arrangement made between European and Mediterranean
television services for the distribution of television programmes.

Eutelsat: See *European Telecommunications Satellite Organization.*

event: (1) In *computer graphics*, an action by the operator , such
as pressing a switch or clicking a mouse button, that produces an
input. **(2)** In *Digital Video Broadcasting*, an event is what would
be considered a programme in analogue transmission terms,
i.e. MPEG-2 video and audio *elementary streams* with common
start and end times. **(3)** In videotape editing, a set of actions
initiated and controlled by the editing computer to be recorded in
one pass.

event queue: A list of input commands in a *computer graphics* system that
have been organized in sequence and are waiting to be processed by the
main operating software.

exchange: (1) A regional centre for the inspection and despatch of film
release prints to individual cinemas. **(2)** A switching centre on a *wide
area network.*

exciter lamp: An incandescent lamp used as a light source for the sound
scanning beam in a photographic sound reproducer .

excursion: The limits of variation of a signal or motion.

execute: In a *computer*, to obey and act on the instructions contained in its
operating *program.*

expander: A device for increasing or restoring the *dynamic range* of an
audio signal.

exponential movements: *Animation* where the speed of movements is
proportional to the size of the field.

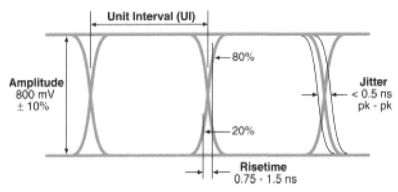

Figure E.3   Eye display for serial digital video at 270   Mbit/s

exposure: The   process of subjecting a photo-sensitive surface to electromagnetic radiation, as in a photographic or video camera.

Exposure Index (EI):   Relates   to the sensitivity of a photographic emulsion. Lower numbers mean less sensitivity . See also *ASA (2)*, *DIN (1)* and *speed*.

exposure meter: An instrument for determining the amount of light falling onto or reflected by a scene to be recorded by a photographic or video camera.

Extended Definition Television (EDTV): A television system with higher horizontal *resolution* images but using the same number of *scan lines* to be compatible with existing 525-line   or 625-line signals. (Hist.)

extra high frequency (EHF): The region of the *electromagnetic spectrum* between 30 GHz and 300 GHz. See **Appendix N**.

extra-high tension (EHT): Describes the extremely high voltages found in *CRT* driving circuits in the region of 25 kilovolts.

extra low frequency (ELF): The region of the *electromagnetic spectrum* between 30 Hz and 300 Hz. See **Appendix N**.

extreme close-up (ECU):   A shot generally showing only a very small subject or part of the subject.

extreme long shot (ELS): A shot showing a very wide general view of the setting, showing vast distances.

eyebrow: Small metal *flag (1)* fixed to the *matte box* of a camera to shade the lens from unwanted light.

eye display: A *waveform* used to evaluate the performance of a digital transmission or *telecommunications* channel. Also called an *eye pattern*. See **Figure E.3**.

eye lights: See *catch lights*.

eye pattern: See *eye display* and **Figure E.3**.

F

F&E: Film and Equipment. A screw-locking video connector originally used in the US film industry . See **Figure F.1**. (Hist.)

Figure F.1   F&E socket

fade: (1) The controlled gradual reduction or increase of a picture image or an audio or video signal. **(2)** The decrease in strength of a received signal due to *transmission (1)* variations.

fade down: The gradual reduction of an audio signal or lamp intensity .

fade-in: See *fade-up*.

fade-out: The gradual disappearance of a picture, usually to uniform black *(fade to black)*, or the decrease of an audio signal.

fade-up: The gradual appearance of a picture, usually from a uniform black, or the increase of an audio signal.   Also called *fade-in*.

fader: A controller used to vary audio signal level or control picture *contrast* to produce *fade (1)* effects.

fader start: A switch incorporated in a *fader* to remotely operate separate equipment.

fade shutter: A two-bladed *shutter (2)* in a motion picture camera, the opening of which can be altered during running to vary the *exposure* for *fade (1)* and *dissolve* effects. In a printer, a variable *aperture (1)* in the light beam for the same purpose.   Also called a *fading shutter*.

fade to black (FTB): See *fade-out*.

fade up: The gradual increase of an audio signal or lamp intensity .

fading shutter: See *fade shutter*.

fairings: Acceleration or deceleration when movement in *animation* starts, stops or alters speed.

fall off: Gradual reduction in *illumination* from the centre of a *screen (1)* to the edges and corners.

fall time: The time taken for the trailing edge of an electrical pulse to fall between two fixed levels, usually from 90 per cent to 10 per cent of its maximum *amplitude* (compare *rise time*).

fan wipe: A *wipe* in the form of an opening fan. See **Figure W.1**.

farad: *SI unit* of electrical capacitance. The capacitance when 1 *coulomb* of charge generates 1 *volt* across a capacitor. Named after the nineteenth century English scientist Michael Faraday .

Federal Communications Commission (FCC): An independent USA government agency established in 1934. It regulates radio, television, telephone and cable communications services in the USA. (W ebsite: www.fcc.gov/oet).

Federation Against Copyright Theft (FACT): A UK or ganization to protect film and video copyright holders.

Fédération Internationale des Archives du Film (FIAF): An international organization based in Paris, France, which keeps data on film archives around the world and or ganizes archivists' meetings.

feed: A television signal from a remote location, e.g. a *satellite (2)* feed.

feedback: A proportion of the output of an amplifying system being returned to the input. See also *negative feedback* and *positive feedback*. See also *howl round* and *threshold howl*.

feedhorn: The assembly at the focus of a *dish (2) aerial* from which *microwave* signals are radiated onto a *parabolic reflector* to form the transmitted *beam*. See **Figure P.1**.

Fernseh-und Kinotechnisehen Geselisehaft (FKGT): German film and television technical society.

ferric oxide tape: Magnetic recording tape using iron oxide particles. It has lower *coercivity* and *remanence* than *chromium dioxide tape*. Also known as *oxide tape*.

ferrite: A magnetic material of high *permeability*, used for the *cores (2)* of high efficiency inductors in *radio frequency* applications.

fettle: A set of colour manipulation controls on *Quantel's* most sophisticated digital *computer graphics* equipment.

fibre channel (FC): A *protocol* for a high capacity *local area network* commonly operating at a data rate of 1 Gbit/s. It can use *shielded twisted pair*, *coax* or *optical fibre* as a physical medium and is often used to interconnect high capacity digital storage devices in a *video server* system.

fibre channel – arbitrated loop (FC-AL): A *fibre channel* network *topology* in the form of a closed dual-loop. No *hub* is needed but only two *nodes* can communicate at one time.

Fibre Distributed Data Interface (FDDI): A standard for a 100 Mbit/s *LAN*, based on *fibre optic* or wired interconnections configured as dual counter rotating *token rings*. This provides a high level of fault tolerance by creating multiple connection paths between *nodes* so that connections can be established even if a ring is broken.

fibre optics: A system using light transmission along a single *optical fibre* or bundle of fibres. It can be used to illuminate enclosed areas, e.g. car dashboards, or as a signalling and high-capacity *telecommunications* system.

field: (1) In television, one half of the picture image consisting of alternate *scan lines*. Two *interlaced* fields are needed to complete a *frame (2)*. See **Figure I.2. (2)** A group of contiguous bytes forming a single piece of data.

field blanking: The part of the *blanking period* between *fields (1)*. See also *vertical interval* and **Appendix L**.

field chart: A plastic sheet, punched for register pegs and engraved with the exact area of the maximum field of view for a certain *set-up (2)*.

field lens: A large diameter *supplementary lens* used to increase light transfer in an optical system, especially in *aerial image* projection.

figure-of-eight microphone: A *bi-directional microphone* with a *polar diagram* shaped like the number eight. See also *directional microphone* and **Figure D.5**.

file: In computing, a named *program* or a collection of data.

file extension: A 1, 2 or 3 character *alphanumeric* code appended to a file name to indicate the type of file.

file server: A special *computer* on a *network* where shared *software* resources are stored. See also *server*.

file transfer pr otocol (FTP): A *protocol* used for exchanging *files* on *TCP/IP* networks including the *Internet*.

fill: (1) The video source in a *keyer* which is inserted into the hole cut by the *key signal* in the *background (1)* video. **(2)** To *load (3)* a *computer* with data (usually a *program*) from disk.

fill light: Light directed to fill shadows, usually directed from the opposite viewpoint to the *key light*.

film: (1) A thin layer, especially light-sensitive photographic emulsion, on a transparent *base*. **(2)** Specifically, motion picture film, a strip of such material, usually perforated along the edge(s) for transport and *synchronization*. **(3)** A general term for the craft and techniques of *cinematography*.

film base: The support upon which the light-sensitive photographic emulsion is coated.

film speed: The sensitivity of a photographic film, as determined by standardized ratings such as *ASA* or *DIN (2)*.

filmstrip: A compilation of individual still photographs, illustrations and/ or text on a single length of film for projection purposes, often associated with lectures or demonstrations.

filter: (1) In lighting, coloured transparent material which modifies the colour of the transmitted light by selective absorption of certain *wavelengths.* See also *gel.* **(2)** Transparent material which transmits visible light but absorbs or reflects *infra-red* radiation. See also *heat filter.* **(3)** An electronic circuit which blocks unwanted frequencies and passes the required ones.

filter pack: In motion picture printing with *subtractive colour* control, the group of colour and *neutral density filters (1)* required to produce the correct *exposure* from a given negative.

finger: A very small rectangular *flag (1)* to control part of the light beam from a *luminaire.*

fire rollers: Pairs of small rollers through which film passes in a *projector,* intended to stop transmission of flames. It is also called a *fire-trap.*

fire trap: See *fire rollers.*

FireWire: A digital *interface (1)* developed by Apple Computers, most commonly used for interfacing between digital video cameras, *desktop editing* systems and disk drives. The first *IEEE*-1394 FireWire standard operates at a maximum data rate of 400 Mbit/s. (See also *iLink.*)

firmware: *Software* stored in a *non-volatile store* such as *ROM.*

first generation: The original recording made on a videotape. A copy of this recording on a different tape would therefore be second generation.

first in first out memory (FIFO): The basic memory structure in a data *buffer (1).*

fishpole: A hand-held *microphone boom,* typically 2 metres in length.

fisheye lens: A lens with intentional visual distortion, giving an angle of view of 180 degrees.

fixture: See *luminaire.*

flag: **(1)** A small opaque surface used in studio lighting to shade a particular area. See *french flag* and also *blade, cutter, dot, eyebrow, finger, target.* **(2)** In data and *telecommunications* systems, a *bit* which can be set, reset or tested by the system.

flagging: Horizontal *jitter (3)* at the top of a television picture, usually resulting from timing errors in the playback signal from a *videotape recorder.*

flanging: An audio effect similar to *phasing* but more intense.

flare: The scatter of light in a lens. It adds undesirable light to the dark areas of an image.

flash: A brief item interrupting a *broadcast* programme.

flash frame: **(1)** In a motion picture negative, a heavily exposed single *frame (1)* producing a clear area in the corresponding positive print. **(2)** In video, a single *frame (2)* of white used to mark a specific point in the count-down to a programme start, usually at minus 5 seconds. **(3)** An unintentional or unrelated film or video frame in an edited programme.

flat: (1) A rectangular covered wooden framework used for building scenery. **(2)** Describes a visual scene with not enough *contrast*.

Fletcher-Munson curves: Equal *loudness* graphs showing how the sensitivity of human hearing varies with frequency in relation to sound level for pure tones.

flick book: A series of small drawings or photographic prints bound together at one edge which give the illusion of motion when flicked through by the user 's thumb. They have been published since the *Kineograph* was patented by a Mr Linnett in 1868. Also called *flicker book* or *flip book*. See also *Mutoscope*.

flicker: Random or regular short period variations of *screen (1)* luminous intensity.

flicker book: See *flick book*.

flies: (1) See *fly tower* (colloq.). **(2)** Hoists or wires hung from the *lighting grid* in a studio and used to raise (fly) scenery or performers upwards.

flip: (1) Momentarily switching from one *projector* to another in a *slide* presentation. **(2)** A video effect that appears to rotate a video picture about a vertical axis.

flip book: See *flick book*.

flip chart: One of a sequence of large pages of text or diagrams usually presented on an easel. When each one has been used, it is 'flipped' over the top of the easel to disclose the next. The term is used colloquially to describe the flip chart supporting easel and pad.

float: (1) A movable part of a studio or stage set which can be easily lifted into or out of position. **(2)** In motion picture projection, a comparatively slow vertical unsteadiness of the image. **(3)** Movement of one image against another in multiple-*exposure rostrum camera* work.

flood: A *luminaire* giving a wide spread of light.

flood track: A photographic sound track exposed to show the uniform maximum width of the image area.

floor manager (FM): The person in the studio production team responsible for operations on the studio floor , cueing artists, etc.

flop-over: A visual effect in which the picture is shown reversed from left to right.

floppy disk: A data storage medium using flexible magnetic *disks*. Also called a *diskette*. It was invented in 1972 by an American, Alun Shugart.

flow chart: A graphical representation of alternative routes available in a *computer program*.

fluid head: A camera *head (4)* for a *tripod*, with *pan* and *tilt* facilities made smooth by fluid *damping*.

fluting: See *edgewave*.

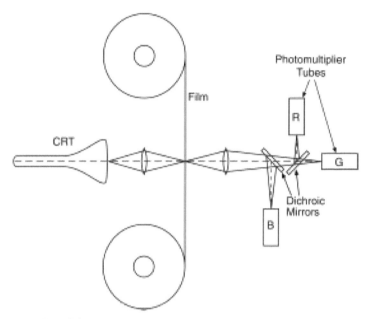

Figure F.2   Flying spot scanner

flutter: Periodic variation in speed of an analogue recording medium, occurring above about 10 times per second. See also *wow*.

flutter echo: An audio ef fect caused by a repeated echo between two facing parallel surfaces.

flux: See *luminous flux* or *magnetic flux*.

flux density: The intensity of a magnetic field. See *magnetic flux*.

flyback: Moving the electron beam of a *CRT*-based television monitor back to the starting point ready for the next line or field scan to begin. Called *retrace* in the USA. See **Figure I.2**.

flying erase: The facility for erasing single *video tracks* in a *helical-scan videotape recorder* by a rotating erase *head (2)*, rather than a fixed one. This is essential to allow accurate *insert editing* on *videotape*.

flying erase head: An erase *head (2)* incorporated in the rotating *drum* of a *videotape recorder*. It is also called a *rotary erase head*.

flying spot scanner: A device for producing video pictures from film based images by *scanning* them with a spot of light, usually produced by a *cathode ray tube* . The transmitted light is divided into its *RGB* components by *dichroic* mirrors and collected by three *photomultipliers* or *photocells* to provide the *component video* signals. See also *telecine* and **Figure F.2**.

flyman: Staff working in the *fly tower* on raising and lowering scenery , etc.

fly tower: An area above a stage from which scenery, etc. can be raised or lowered, hidden from the audience. Also called the *flies (1)*.

FM control: See *variable-frequency control*.

f-number: Indicates the brightness of the image formed by a lens. It is the *focal length* of a lens divided by its effective *aperture (1)*. This assumes that lenses transmit all the light falling on them. Dif ferent lenses may transmit different amounts of light for the same f-number , so may not giving equal image brightness (compare *T-number*). Also called *f-stop*.

FOCAL: An international professional trade association formed in 1985 to represent commercial film/ *audio-visual* libraries, professional film researchers, producers and others working in the industry . (Website: www.focalint.org).

focal length: In a *compound lens*, the distance from the secondary *principal point* of the lens to the image *focal point*. See **Figure P.10**.

focal plane: The plane through the *focal point* of a lens at right angles to its *optical axis*.

focal point: The point behind a lens at which light rays from an object at infinity form the most sharply defined image. More correctly called the image focal point, since there is also an object focal point on the object side of the lens. See **Figure P.10**.

focus: (1) Adjusting the lens of a camera or *projector*, nearer to or further from the plane of the gate or *aperture (2)*, to make the image sharply defined. Hence, the position at which the sharpest image is formed. **(2)** The point in a *satellite (2)* or *microwave dish aerial* at which the signal is at maximum strength.

fog: Additional *density* in a processed photographic image caused by accidental *exposure* to light, radiation or a chemical fogging agent.

fog level: The minimum *density* of the unexposed area of processed film.

foil: A sheet or roll of thin transparent or translucent material, either plain or carrying prepared images, as used on the *platen (2)* of an *overhead projector*.

foldback: A cueing signal sent to musicians enabling them to hear a previously recorded performance, or that of another musician playing at the same time, or indeed themselves.

Foley effects: The footsteps, door creaks, leather squeaks, etc. which contribute to the identity and character of a programme or film. Named after an American, Jack Foley.

follow focus: A camera technique in motion picture and *rostrum camera* photography to compensate for varying subject distance to maintain image *focus*.

follow spot: A powerful spotlight used to follow the movement of an artiste.

font: Originally, a set of printers ' type of uniform style and size, more correctly called a *fount*. Now also refers to a uniform style of *alphanumeric* characters available for use in a video or *computer graphics* system.

foot: A measure of motion picture film length representing a specific number of *frames (1)*: 16 for 35 mm, 40 for 16 mm and 72 for Super – 8.

footage numbers: See *edge numbers*.

foot-candle: Unit of illumination, a measure of the light falling on an object. Equal to the illumination from a source of one *candela* at a distance of one foot. (Hist.)

foot-lambert: Unit of surface *brightness*, reflected light intensity or *luminance (1)*. Equal to the emission or reflection of one *lumen* per square foot. (Hist.)

footprint: The area on the surface of the Earth within which a *satellite (2) transmission* can be received.

forced development: Motion picture processing using increased time or temperature to compensate for under-*exposure* of the original film in the camera.

foreground: In a *keyer*, another name for the *fill (1)* or *insert (3)* video used in the 'hole' cut by the *key signal* in the *background (1)* image.

form feed (FF): In word-processing, a character which causes the print or display position to move to the start of the next page, the length of the page having been defined.

format: (1) The style or method of presentation of an image or its physical dimensions. **(2)** The configuration of information for display , transmission, storage or retrieval. **(3)** To prepare a storage medium, e.g. a *floppy disk*, so that it can record data.

forward error correction (FEC): Adding *error correction codes* to a *packetized* data stream to be transmitted or recorded so that the receiver or replay system can check the reliability of the data and try to correct any errors which have occurred. The system is used when there is no return path to allow the data to be retransmitted (compare *automatic repeat request (ARQ)*).

fount: See *font*.

fovea: The area in the *human eye* on the *retina* directly behind the *lens (3)* where the majority of *cones* are concentrated. See **Figure H.6**.

frame: (1) An individual picture image on motion picture film. **(2)** In video, the picture formed by a pair of *interlaced fields* or in one sweep by *progressive scanning*. **(3)** In *computer graphics*, one *refresh* of a display image. **(4)** In data communications, a variable length group of data *bits* or *bytes* used as the basic unit for transmission. **(5)** In *MPEG* audio data compression, the length of the basic group of samples on which the *psychoacoustic* analysis is performed. It varies in duration from 8 milliseconds to 12 milliseconds, depending on the *sampling frequency.* **(6)** To adjust the elements of a shot in a camera *viewfinder (1, 2)* to give the desired aesthetically pleasing ef fect. **(7)** In presentations, a holder for *overhead projector* foils. **(8)** See *slide mount.* **(9)** A mechanical assembly which can be used to suspend a *luminaire.* Also called a *yoke (2).*

frame buffer: See *frame store.*

frame count cueing (FCC): In motion picture printing, a control system based on electronically counting the number of *frames (1)* passing through the machine.

frame counter: In motion picture equipment, a *counter* registering the number of *frames (1)* of film passing through it.

frame grab: In *computer graphics*, capturing a video image via a *frame buffer* for later manipulation by a graphics device.

frame line: The space between consecutive *frames (1)* on photographic film.

frame rate: The number of pictures presented or recorded each second. It is measured in *frames per second.*

frame roll: A vertical roll of a television image usually due to a fault in the receiver.

frames per second (fps): For sound film a rate of 24 fps has been standard since 1927. An effective doubling to 48 fps is used for improved quality of film presentation by projecting each image twice. The *IMAX* system shoots film at 48 fps. In television and video 25 fps is used with 625-line systems and 29.97 fps or 30 fps is used with 525-line systems.The ATSC digital television system includes *frame rates* of 24, 25, 29.97, 30, 59.94 or 60 fps.

frame store: A memory capable of storing a complete *frame (2)* of video information in digital form. Used for television *synchronizers (2)*, *standards conversion* and sophisticated graphics and special ef fects units. Also called a *frame buffer.*

framing: (1) Adjusting the film image in the *projector* gate *aperture (2)* so that the projected image appears centred in the vertical dimension of the *screen (1).* **(2)** Adjusting the elements of a shot in a camera *viewfinder (1, 2)* to give the desired aesthetically pleasing ef fect.

free field: A sound field with no reflective surfaces.

free head: A recording *head (2)* without electronics, provided on an audio *cassette* recorder for use with *tape-slide* control equipment.

freeze: (1) See *lock-up.* **(2)** See *freeze frame.*

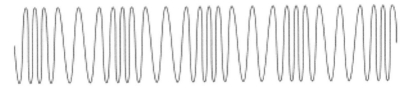

Figure F.3 Frequency modulation

freeze frame: A visual effect in which a single picture from a moving scene is repeated to hold the action stationary for as long as is required. Also called *hold frame* or *still frame* or just *freeze.*

freeze grab: A facility in a video field or *frame store* to monitor a continuous input signal and select a field or *frame (2)* at a predetermined rate, reproducing this as a *still frame* until updated by the next.

french brace: A strut used to support scenery .

french flag: A device to block stray light from entering the camera lens. See also *flag (1).*

frequency: The number of repetitive cycles of a periodic signal in one second. Its numeric value is given in *hertz* (Hz).

frequency band: A part of the *electromagnetic spectrum* allocated for a specific purpose, e.g. the *UHF* band for terrestrial television broadcasts.

frequency modulation (FM): Carrying information by varying the frequency of a *carrier signal* value while keeping its amplitude constant. It was invented in 1933 by Edwin Howard Armstrong, an American electrical engineer. See **Figure F.3.**

frequency response: The relationship between the *gain* of a device and *frequency.*

frequently asked questions (F AQ): A document, often designed as a *hypertext* document, which answers questions on a technical topic.

Fresnel lens: A lens formed by concentric stepped rings, each of which is a section of a convex surface. Often used as a *condenser lens* in an *overhead projector* or spotlight. Named after A. J. Fresnel, the nineteenth century French physicist. See **Figure F.4.**

Fresnel zones: Areas of alternate destructive and constructive interference between two *microwave aerials* on a *radio link.* The zones form ellipsoidal shapes between the *aerials.* See **Figure F.5.**

fringing: (1) False colours around an image arising from a defective or maladjusted optical system. (2) The effect of using a full track test tape on a multi-track tape machine which results in an exaggerated low-frequency response.

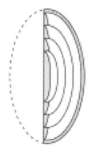

Figure F.4 Fresnel lens

front axial projection: A trick effects shot, in which the *background (3)* to the action is provided by projecting an image along the axis of the camera lens on to a highly directional beaded reflecting *screen (1)*. Also known as *reflex projection*.

front end: In motion picture production, front end operations are those concerned with the original photography , sound recording, *editing* and preparatory work leading up to the first presentation.

front feed: Type of microwave *aerial* in the form of a *parabolic reflector* with the *primary radiator* mounted at the focus (compare *offset feed*). See **Figure P.1**.

front of house lights (FOH): *Luminaires* mounted above an audience seating area, usually on *barrels* and often concealed.

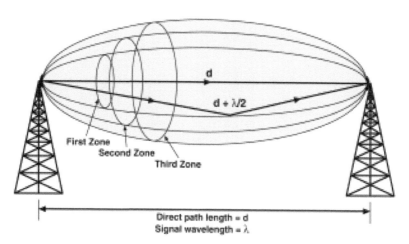

Figure F.5 Fresnel zones

front porch: In a television waveform, the brief period at *blanking level* between the end of picture information and the start of the horizontal *sync pulse (2)*. See **Appendix K**.

front projection (FP): The presentation of a picture to be viewed from the same side of the *screen (1)* as the *projector*.

f-stop: See *f-number*.

full bleed: A full-page illustration which runs of f the page at all or some of its edges.

full-duplex: A mode of data transmission in which two *stations* can communicate in both directions simultaneously .

full-field: covering the complete *active picture* area of a video picture.

fusion point: In *computer graphics*, the point at which the rate of *refresh* redrawing the display makes it appear steady .

fusion splice: A means of joining *optical fibres*.

G

gaffer: The senior lighting electrician in a film or television studio.

gain: (1) A ratio of amplification or attenuation, usually expressed in *decibels*. **(2)** The measure of the reflectivity of a surface compared with a perfect matt white diffusing surface under identical conditions. **(3)** The measure of the performance of an *aerial* system compared with a simple *dipole*.

gallery: The television studio production control room where the director and production team control and produce the current programme being made.

gamma: (1) A measure of photographic film contrast, the gradient of the straight-line portion of the *characteristic curve* relating *exposure* and *density*. See **Figure C.2**. **(2)** In television, the relationship between the original scene brightness and the corresponding displayed brightness. For a camera it is the slope of the curve relating the logarithms of the incident light and the output voltage. Similarly for a display device, the input voltage and the light output.

gamma correction: In television, a non-linear amplification is applied in all picture source equipment to ensure satisfactory reproduction on display devices which have a non-linear response. See **Figure G.1**.

gamma rays: Electromagnetic radiation with wavelengths shorter than about 0.1 picometres.

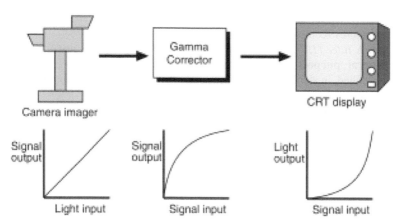

Figure G.1 Gamma correction in television

gamut: The range of voltages allowed for a video signal, or a component of a video signal. Voltages outside this range may lead to clipping when the signal is prevented from going beyond the allowed range.

ganged: Controls which are mechanically connected so that movement of one causes movement of the others.

gap: The separation between the pole pieces of a magnetic recording *head (2)* at right angles to the direction of motion of the tape. The width of the gap determines the shortest *wavelength* that can be replayed from the tape. Also called *head gap.*

garbage in, garbage out (GIGO): A colloquial maxim in computing.

gas-discharge lamp: A light source in which an electrical discharge takes place in a rare gas within an *envelope (1)* to provide the *illumination.*

gate: (1) In photographic equipment, the *aperture (2)* at which motion picture film is exposed or projected. **(2)** The assembly that supports and positions the *slide* in a *slide projector.*

gate temperature: In an optical projection system, the temperature reached at the *aperture (2).*

gateway: A *computer* which acts as a point of connection between different *networks.*

gauge: The width of photographic film or *magnetic tape,* usually expressed in millimetres.

gauss: Unit of magnetic induction. It is the flux density which exerts a force of one *dyne* per centimetre on a wire carrying ten amperes perpendicular to the field and is named after the nineteenth century German mathematician K. F. Gauss. The preferred *SI unit* is the *tesla,* where 10 000 gauss = 1 *tesla.*

gauze: Transparent or translucent material used to create special visual effects.

geared head: A camera *head (4)* for a *tripod,* with geared *pan* and *tilt* features (compare *fluid head*).

gel: A *filter (1)* for lighting which changes the colour or quality of light.

general purpose interface bus (GPIB): A digital *parallel interface* defined in standard *IEEE 488* and often used to control laboratory instruments. Also known as HP-IB or IEC-625.

general view (G/V): An establishing shot in film or video.

generation loss: Degradation of analogue video and sound quality caused by successive stages of transfer or *dubbing.*

geneva movement: A mechanical *intermittent motion* device based on a *maltese cross* shape, interacting with a pin rotating on a disc. See **Figure G.2**.

genlock: Connecting a video reference from a remote source, often *colour black,* to a local *sync pulse generator* to provide full *synchronization.* Sometimes called *backtiming (2)* in the USA.

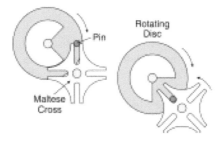

Figure G.2   Geneva movement

geostationary arc: That portion of the *geostationary Earth orbit* from which signals can be received at a given point on the Earth.

geostationary Earth orbit (GEO): The unique circular orbit 35 785 km (22 237 miles) above the equator in which a *satellite (2)* is effectively stationary relative to a point on the Earth. It is the special case of a *geosynchronous orbit* where the *satellite (2)* is constrained by *stationkeeping* from moving outside its allocated position. See **Figure O.4.**

geosynchronous orbit: A circular orbit at a distance of about 42 000 km from the centre of the Earth where a *satellite (2)* has a period exactly equal to the rotational period of the Earth (a *sidereal day*). Gravitational influences induce north and south movements of the *satellite (2)* unless *stationkeeping* is used to make it geostationary.

getter: A substance used to remove, by chemical combination, unwanted gases in valves, *cathode ray tubes*, etc.

ghosting: Shadow or weak double images appearing in a picture. In television can be caused by unwanted reflections in a cable system or from physical objects in the path of a radiated signal. In film projection it can result from an out-of-phase *shutter (2)*.

gigabit Ethernet: An extension to the original *Ethernet* data network standard which operates at a data rate of 1 gigabit/sec.

gigabyte: 1024 *megabytes*, equal to 1 073 741 824 bytes.

gigahertz (GHz): 1000 megahertz. See *hertz*.

gliding frequency: (1) An audio response test using a continuous slowly changing frequency, usually from 20 Hz to 20 kHz. **(2)** A form of *tape-slide* control using *frequency modulation*.

glitch: An intermittent error in a circuit.

global beam: A *satellite (2)* transmission covering the entire surface of the Earth which is visible from the *satellite (2)*, corresponding to about one third of the Earth's surface.

global positioning system (GPS): A *satellite (2)* based navigation system.

global system mobile (GSM): The second generation of mobile telephone technology which allows mobile telephones to carry data and, with the use of *wireless application protocol (WAP)*, basic e-mail and *Internet* services.

glossy: Describes a surface which reflects light predominantly in one direction, e.g. a gloss painted object.

gnats: Abbreviation for gnats whisker. A very small technical adjustment. See also *smidgen.* (Colloq.)

gobo: (1) A thin metal plate cut and mounted in front of a *luminaire* to provide a desired pattern of shadows on a studio backing. **(2)** A portable wooden or cloth screen, usually stand mounted, to act as a light shield between a lighting unit and camera lens.

gofer: A *runner.* (Colloq.)

golfball: A type of printer or typewriter with the printing characters embossed on a small sphere. (Hist.)

graded index fibre: *Optical fibre* in which the *core (2)* has a smoothly changing *refractive index* outwards towards the *cladding* (compare *stepped index fibre*).

grades: In television picture monitors, grade 1 is *broadcast standard,* grade 2 is general purpose and grade 3 is no better than a converted domestic television *receiver.*

grading: (1) In motion picture work, the selection of the colour and *density* printing values required for each scene of an assembled roll of negative film. **(2)** In video *editing,* matching the *colour balance* between shots.

grain: The physical structure of a processed photographic image. It is not normally seen as such, but as clumps of grains, for which the term *graininess* is more appropriate.

graininess: The visibility of clumps of grains in a processed photographic image.

gramophone: An *audio disc* player. (Hist.)

granularity: The objective measurement of *graininess.*

graphical user interface (GUI): A method of interacting with a computer using windows and a pointer controlled by a *mouse, tracker ball, joystick* or *graphic tablet* and pen. See also *WIMP.*

graphic equalizer: A multi-section audio *filter (3)* with slider level controls at *octave* or third *octave* intervals which provide adjustment and indication of the shape of the *frequency response* characteristic.

graphics: (1) Artwork, *captions*, lettering, photographs, etc. used in programmes. **(2)** Output presented in graphical form as a video signal from electronic graphics equipment.

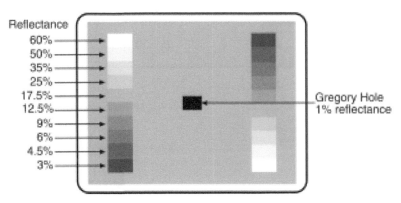

Figure G.3   Grey scale chart

graphic tablet: A device for inputting graphical information to a *computer*. See also *data tablet*.

graticule: A series of lines engraved or displayed on the face of a display device, or engraved on a separate transparent sheet fitted in front of a display device, used particularly for the measurement or location of an image.

gravure coating: A method of coating *magnetic tape* similar to the gravure printing process.

Green Book: The standard for *Compact Disc – Interactive (CD-i)* .

green print: A   processed motion picture film having an excess humidity in the emulsion, which may cause unsteadiness or damage in projection.

Greenwich Mean Time (GMT): The mean solar time on the Greenwich meridian, used as the international basis for time calculations. Now known as *Universal Time Co-ordinated.* (Hist.)

grey scale: A test chart with incremental   *brightness* steps from black through grey to white.   The steps may conform to a logarithmic relationship. See **Figure G.3**.

grid: (1) Uniformly spaced points or lines in two or three dimensions within which an object may be defined. **(2)** The current control electrode of a valve, *cathode ray tube*, or camera *tube (2)*. **(3)** The framework above a studio floor for suspending scenery and lights.

grille: See *cross hatch*.

grip: A general member of a film crew who transports equipment, lays camera rail tracks or pushes a camera.

ground: An electrical zero-voltage reference, i.e. at   *earth* potential or connected to *earth*.

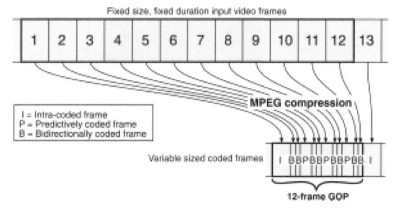

Figure G.4 A 12-frame GOP (group of pictures)

grounding: Electrically connecting to *earth*.

ground loop: See *earth loop*.

groundrow lights: Lighting units on a stage or studio floor for lighting a *cyclorama* or scenery. Also called *cyc lights*.

ground station: See *earth station*.

group: (1) The ganging together of a selection of circuits on a lighting desk or audio mixer. **(2)** Controls on a mixing console or lighting desk simultaneously affecting a selected set of channels.

group delay: An effect in a transmission system where some frequencies travel faster than others (not the same as *phase shift*).

group of pictures (GOP): A sequence of *MPEG* compressed *frames (2)* whose length is the number of *frames (2)* between successive *I-frames*. See **Figure G.4**.

G-spool: A *spool* holding 5 minutes of *videotape*, used for commercials. (Hist.)

guard band: (1) A blank region left between recorded tracks on *magnetic tape* in order to avoid *crosstalk*. **(2)** Unused *radio frequency* spectrum between radio channels.

guide: In *magnetic tape* recorders, a pillar or roller used to direct the tape along the *tape path*.

guide path: See *tape path*.

guide track: A speech track recorded where *background (2)* is high, to serve as a guide for the actor to repeat the dialogue again in a studio during *post production*.

Guild of British Camera Technicians (GBCT): Association of British camera technicians worldwide, which aims to improve standards and promote excellence in film making. Its members are elected by their peers and their details are represented in an on-line database. (W ebsite: www.gbct.org).

Guild of Television Cameramen (GTC): The GTC is not a trade association but aims to ensure and preserve the professional status of television cameramen in the UK. (W ebsite: www.gtc.org.uk).

gun microphone: A very directional *microphone*, also called a *rifle microphone.*

gutters: The dark bands between the image areas of a compartmented multi-*screen*. See *egg boxing*.

gyroscopic error: A timing error impressed into a *videotape* recording by the physical movement of the recorder while making a recording.

H

hack: A journalist. (Colloq.)

hair in the gate: Images of hair-shaped particles seen on the *frame line* of motion picture images after processing.

hairpin: US term for *U-link*.

halation: Unwanted *exposure* surrounding the photographic image of a bright object, caused by light scattered within the emulsion layer or reflected from the base of the film.

half-duplex: A mode of data transmission in which each *station* can send to the other, but only in one direction at a time.

half frame: A photographic film frame with smaller dimensions than the standard (35 mm) *frame (1)*, i.e. 24 × 18 mm instead of 24 × 36 mm for *slides* or 21 × 8 mm rather than 21 × 16 mm for *cinematography*.

half-tone: Tonal differentiation in an image by a series of ruled, etched or pigmented lines or vignetted dots evenly spaced at specific intervals.

half track: A magnetic recording with two separate tracks on the same tape. Half track *heads (2)* are narrower than stereo *heads (2)* and have a wider *guard band* between them.

Hall effect: A voltage change across a semiconductor material that has current passing through it. This results from a change in magnetic field at right angles to the direction of current flow . It is used in some keyboards and proximity detectors. Named after the American physicist Edwin H. Hall.

halo: A lens *aberration* giving a circular *flare* around a point image.

hand control: A manual remote control for operating an automatic presentation device.

hand-held: Using a camera without a *tripod* mount.

hand-shake: A two-way exchange between two *computers* or a *computer* and a *peripheral*, e.g. to enable a printer to signal if it is ready or not ready to receive data.

Hanover bars: A *PAL* system defect produced by a *simple PAL* decoder when severe *phase (1)* errors occur in the signal. Adjacent pairs of *scanning lines* have a noticeably dif ferent colour and appear to move up or down in the picture. It is sometimes called *venetian blinding*.

hard copy: A printed, thermally or electrically produced print-out as opposed to the image on a computer monitor screen.

hard cut: In *slide* projection, a very fast *cut*, as rapid as *lamp inertia* will permit.

Figure H.1   Hard disk drive assembly

hard disk drive (HDD):   A digital magnetic data storage device comprising one or more rigid *platters (2)*. See also *block (3)*, *cylinder*, *sector*, *surface*, *track (3)* and **Figure H.1** and **Figure H.2**.

hard edge: A sharply defined boundary of a picture image area, as in a *matte* or *wipe*.

hardware: (1) The physical components, integrated circuits, etc. which, together with the *software (1)*, make up a *computer* system. **(2)** The picture and sound equipment used for presentation, as opposed to the programme *software (2)*.

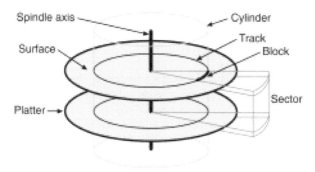

Figure H.2   Hard disk drive terminology

Figure H.3   Head drum assembly

harmonic: A frequency which is a whole multiple of the fundamental frequency.

harmonica connector: A multi-pole strip connector resembling a harmonica from the holes in its construction.   Also called a *chocolate block.*

harmonic distortion: Adding a different frequency component not present in the original signal.

harmonizer: An electronic *pitch changer* to raise or lower   *pitch (2)* without altering the time aspects of a programme.

hash marks: Unofficial *cue marks* scratched on a film *release print.*

head: (1) A general term for   the essential mechanism of a recorder or reproducer, as in ' *projector* head'. (2) In magnetic recording, an electromagnetic *transducer* converting magnetism to electrical signals and vice versa. (3) In photographic sound reproduction, the sound head converts the light passing through the sound track on the film into an electrical signal. (4) The adjustable mounting for a camera on its *tripod.* See also *fluid head.* (5) The beginning of a *roll (2)* of film or tape.

head banding: See *banding.* (Hist.)

head clog: See *clogging.*

head drum: The mechanical assembly containing the video *heads (2)* in a *helical-scan videotape recorder.* Also called a *scanner (1).* See **Figure H.3**.

head end: The originating point on a *cable television* or *aerial* system.

header: The information at the start of a data *packet* containing details of the contents of the *packet* which may include some of the following: source address, destination address, path identification, packet length, data type or *packet* number in a sequence.

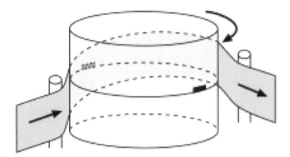

Figure H.4 Helical scan

head demagnetizer: A device producing an alternating *magnetic flux* at a probe. It is used for demagnetizing magnetic recording *heads (2)* and *guides*.

head gap: See *gap*.

head out: Film or tape wound on a *reel* so that it is at the start, ready for immediate projection.

headphones: Small sound reproducers used over , or in the ears, usually excluding extraneous sounds.

headroom: (1) The range of signal levels above the normal working level before *overload (1) distortion* occurs. (2) In *framing (2)* a picture, the space between the top of a subject's head and the top of the picture *safe area*.

headset: *Headphone/microphone* combination.

head wheel: See *head drum*.

heat filter: A *filter (2)* which reduces the heating effect of a light beam by absorbing or reflecting *infra-red* radiation while transmitting visible light.

heat sink: A device, usually metal and often finned, which dissipates unwanted heat by radiation and/or convection.

helical scan: A system of magnetic recording where thin *tracks* are laid down diagonally across a *magnetic tape* during recording by passing the tape in a spiral path around a rotating *drum (2)* containing the magnetic *heads (2)* (compare *transverse scan*). See also *wrap* and **Figure H.4**.

henry (H): *SI unit* of inductance, equal to the inductance of a circuit in which an *electromotive force* of one *volt* is produced by a current of one *ampere* per second. Named after the nineteenth century American physicist Joseph Henry.

heron: A wheeled hydraulic camera *crane (1)* where the camera operator has foot pedals to control the *jib's* elevation and camera platform rotation.

hertz (Hz)

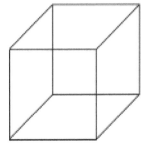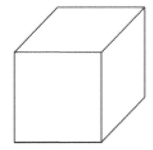

Figure H.5   Hidden lines

hertz (Hz): *SI unit* of *frequency*, named after the nineteenth century German physicist Heinrich Rudolf Hertz.

heterodyne: The production of a beat frequency from the mixing of two alternating signals of which it is the combination of the sum and difference frequencies. The technique is used in radio and television reception equipment.

hexadecimal: A method of expressing each of the 16 possible states of 4 binary *bits* as a single character (0, 1, 2, 3, 4, 5, 6, 7, 8, 9,   A, B, C, D, E or F). Eight-bit *bytes* are dealt with as two four -bit *nibbles*, e.g. 167 decimal = A7 hex = 10100111 binary.

HI arc: See *high intensity arc.*

hidden lines: In *computer graphics*, line segments which should not be visible to the viewer of a three-dimensional display item because they are behind other parts of the same or other items. Similarly *hidden objects, hidden surfaces.* See **Figure H.5**.

high-band: A *videotape* recording system producing *broadcast standard* pictures using a higher frequency *carrier signal.*

High Definition Common Image Format (HD-CIF):   See *CIF.*

High Definition – D5 (HD-D5): High *definition* version of the *D5* format using about 4 :1 video data compression.

High Definition Serial Digital Interface (HD-SDI):   High *definition* version of the *Serial Digital Interface* , defined in standard SMPTE 292M (see **Appendix U**).

High Definition Television (HDTV): A television system having about double the vertical and horizontal *resolution* of conventional 625-line and 525-line systems, together with a *widescreen aspect ratio (1).*

high frequency (HF): (1) In *radio frequency* terms, the region of the *electromagnetic spectrum* between 3 MHz and 30 MHz (see **Appendix N**). **(2)** In audio terms, in the upper part of the *audio frequency* range.

high intensity arc (HI arc): A carbon arc lamp operating at a high current.

high key: Characteristic of a scene in which the majority of tones in the subject or its reproduction are at the light end of the scale.

high-level language: A set of English-like words, symbols, etc., and the rules by which they can be combined to form complex *computer programs*. Some examples are BASIC, COBOL, FOR TRAN, Java and PASCAL.

highlight: The brightest part of a scene or its reproduction.

high-pass filter (HPF): An electrical *filter (3)* which passes high frequencies and attenuates low frequencies.

high-power amplifier (HPA): The high-power section of a *microwave transmitter*, e.g. in a *satellite (2) transponder* or at an *Earth station uplink* site.

hi-hat: See *top hat.*

hither plane: In a *computer graphics* display, the front *clipping* plane defining the limit of three-dimensional space nearest to the viewing plane (compare *yon plane*).

Hi-Vision: The Japanese analogue *high definition television* system developed by *NHK* which uses 1125-lines with 59.94 *interlaced* fields per second and 16 :9 *aspect ratio (1)* . Defined in standards SMPTE 240M and SMPTE 260M (see **Appendix U**).

HMI: A double-ended *metal-halide lamp* with a *colour temperature* of 5600K. (Trade name.)

hold frame: See *freeze frame.*

holography: The recording and representation of three-dimensional objects by *interference* patterns formed between a beam of *coherent* light and its reflection from the subject. See also *laser.*

home: At the base or starting position. Rotary *slide projector magazines (2)* with a zero position usually home to zero.

homing: Term used with automatic *slide projectors* for returning the *magazines (2)* to the start or *'home'* position. Circular ones may be homed by the shortest route.

hop: (1) Periodic vertical unsteadiness of a projected film image (compare *weave*). **(2)** A single transmission path from the Earth to a *satellite (2)* and back to the Earth.

horizontal blanking period: The period between the end of one displayed line and the start of the next during which the *scanning* spot in a *cathode ray tube* -based picture display device returns from right to left. Also called *line blanking.* See also *blanking period* and **Appendix K.**

horizontal polarization: See *linear polarization.*

horizontal split wipe: See *barndoors wipe* and **Figure W.1.**

horn: The acoustic portion of a sound reinforcement *loudspeaker* in which the circular or rectangular cross-section increases in area towards the output end.

horse: A simple support with a spindle for a roll of film wound on a *core (1)*.

hot shoe: A still-camera accessory socket with an integral *shutter (1)* contact connection for a flash gun.

hot spot: An area of excessive local *brightness*, particularly as part of a projected image or illuminated scene.

housekeeping: Looking after the details of automatic presentation in a cinema, such as operating *houselights, tabs,* etc.

houselights: Lights illuminating the audience seating area or general lighting in a studio or theatre.

howl-round: Acoustic *feedback* produced by sound from a *loudspeaker* re-entering a *microphone*. Also called *threshold howl*.

hub: The cylindrical centre of a film or tape *spool*. The term is sometimes used to mean a *core (1)*.

hue: The attribute by which a colour may be identified within the *visible spectrum*, e.g. red is a hue. A hue control may be found on some television *receivers*. (It is called *tint (2)* in the USA.)

hue control: (1) A control found on some television *receivers* to adjust the colours being displayed. It may be called a *tint* control in the USA. (2) A control on a *vision mixer* to select the colour being used in a *chromakey*.

Huffman coding: A digital data compression technique in which more frequently occurring input sequences are represented by shorter codes, and less common sequences are replaced by longer codes. The codes are stored in a Huffman table. See also *variable length coding*.

hum: An audible continuous sound of low frequency, often corresponding to mains frequency of 50 or 60 Hz, or its harmonics. See also *mains hum*.

human eye: The human organ of sight. See **Figure H.6**.

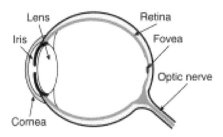

Figure H.6 Human eye

hum bars: Horizontal bars moving slowly up or down the screen on a television picture monitor caused by the injection of unwanted *mains hum* into the video signal.

hunting: **(1)** A cyclic error in a *servo system.* See also *damping.* **(2)** In *videotape* reproduction or film projection, a *low-frequency* instability of picture or sound caused by cyclic variations in tape or film *transport* velocity.

hybrid fibre-coax (HFC): A common physical *network (1)* structure used in *cable television* distribution systems. *Optical fibre* cables are used for the main distribution to street-corner cabinets where the optical signal is converted to an electrical one. *Coaxial cable,* usually in a common protective sheath with *twisted pair* telephone connections, is then used to connect individual subscriber homes or small businesses to the network.

hypercardioid microphone: A *directional microphone* with a *polar diagram* between that of a *cardioid microphone* and a *figure-of-eight microphone.* See **Figure D.5.**

hypersensitizing: Increasing the sensitivity of a photographic film before camera *exposure,* either by chemical means or by brief controlled pre-*exposure* to light.

hypertext: **(1)** Originally a *database* system in which various types of object (such as text, pictures, sound, video and *computer programs*) could be linked to each other . **(2)** The concept of using interactive text links within a document (called hypertext links) to point to and interconnect with other parts of the same document, dif ferent related documents or other resources such as *websites,* images or sound files.

HyperText Markup Language (HTML): An *authoring* language used to create documents on the *World Wide Web.*

HyperText Transport Protocol (HTTP): The protocol used by *clients* and *servers* to communicate on the *World Wide Web.*

hypo: Sodium thiosulphate used as a photographic fixer .

I

icon: In *computers* and in equipment labelling, a symbol clearly representing purpose or function.

I-frame: Intra-coded frame. An *MPEG* compressed picture in a video sequence which takes no account of other frames. It effectively uses *JPEG* compression only and forms the first picture in an *MPEG GOP*. See **Figure G.5**.

igniter: A device to strike a xenon or mercury arc.

iLink: The Sony implementation of *IEEE* 1394 operating at 100 Mbit/s. See also *FireWire*. (Trade name.)

image bit map: The digital representation of a graphics display image as a pattern of bits, each bit representing one or more *pixels*.

image enhancer: A device for improving video picture quality, especially in apparent *resolution* or tonal reproduction.

image intensifier: A device which enables a camera to operate under extremely low light levels, typically starlight.

Image Orthicon (IO): An early monochrome television camera *tube (2)*. (Hist.)

image plane: The plane behind a lens in a camera where the image is formed, i.e. where the *CCD imagers* or photographic film are located. See **Figure D.3**.

IMAX: A *widescreen* motion picture system using 70mm film running horizontally with a 15-perforation interval, giving a *frame (1)* size of 70 × 52.5 mm. (Trade name.)

impairment scale: A scale of subjective assessment of the loss of quality observed in the reproduction of sound or picture, both film and video. See **Table I.1**.

Table I.1 Subjective impairment scale

Degree of Impairment	Scale	Quality Rating
Imperceptible	5	Excellent
Barely perceptible	4	Good
Perceptible, slightly objectionable	3	Fair
Objectionable	2	Poor
Unacceptable	1	Bad

impedance: Resistance to alternating current, measured in *ohms*. It will vary with frequency.

impulse: A mechanical force or electrical signal operating for a very short time.

in-betweening: In *animation*, generating the intermediate frames between two *key frames*.

inches per second (ips): An imperial measurement of tape speed, i.e. audio *cassette* tape at 1.875 ips, *reel-to-reel* at 7.5 ips or 15 ips. (Hist.)

inclination: The angle between the orbital plane of a *satellite (2)* and the equatorial plane of the Earth.

inclined orbit: A *satellite (2)* orbit nominally above the equator but displaced northwards and southwards over its orbital period mainly due to the tilt of the Earth's axis and the gravitational pull of the sun *Satellites (2)* in inclined orbits use tracking *earth stations*, unlike those in *geostationary Earth orbit* which can use fixed transmit and receive *dishes*.

Incorporated Society of British Advertisers (ISBA): UK society founded in 1900 to protect advertisers. It is the only body solely representing the users of marketing communications and actively represents advertisers' interests in key industry bodies. (W ebsite: www.isba.org.uk).

Independent Broadcasting Authority (IBA): The statutory body which was responsible for regulating and transmitting UK *Independent Television* and *Independent Local Radio* until 1991 when it was abolished and its responsibilities split between the *Independent Television Commission* as regulator and *National Transcommunications Limited*, set up as a transmission company . (Hist.)

Independent Television (ITV): (1) The UK independent *broadcast* network. See also *ITV Network Limited*. (Website: www.itv.co.uk). **(2)** Canadian independent *broadcast* organization. (Website: www.itv.ca/index.htm).

Independent Television Association: See *ITV Network Limited*. (Hist.)

Independent Television Commission (ITC): The public body responsible for licensing and regulating television and teletext services in the UK, except for BBC licence-fee funded channels and S4C. It covers terrestrial, *satellite (2)* and cable transmissions and took over these responsibilities from the *Independent Broadcasting Authority* and the Cable Authority in 1991. (Website: www.itc.org.uk).

Independent Television Facilities Centre (ITFC): Formed in 1975 as a technical facility, specifically serving *Independent Television* in the UK, the ITFC now provides services for ITV and other broadcasters and producers. These include technical quality assessment, video facilities, film facilities, content assessment, film/tape storage, materials management, subtitling and audio description. (W ebsite: www.itfc.com).

Independent Television News (ITN): (1) UK independent commercial public service broadcast organization producing news and other factual programmes for national and international channels. (Website: www.itn.co.uk). **(2)** Independent television news provider in Sri Lanka.

index of refraction: US term for *refractive index*.

Indian Ocean Region (IOR): The area of the Earth centred on the Indian Ocean covered by a range of *satellites (2)* in *geostationary Earth orbit*.

Industrial Telecommunications Association (ITA): USA national trade association for users of two-way radio systems, communications equipment dealers and service providers founded in 1953. It is the US telecommunications industry voice before Congress and the *Federal Communications Commission*. (Website: www.ita-relay.com).

information technology (IT): Technology which deals with the storage, processing and distribution of information, usually using computers.

infra-red (IR): Electromagnetic radiation in the region of the *electromagnetic spectrum* extending from beyond the deep red end of the *visible spectrum*, above 700 nm, to the *microwave* region. Generally discernible as heat. See **Appendix N**.

initialize: To get a system into its correct starting state.

ink-jet printer: A printer using a *computer*-controlled electrostatically directed stream of ink droplets.

inkie: An incandescent tungsten or *tungsten-halogen lamp*, as distinct from a carbon arc or *metal-halide lamps*. Also called an *inky*.

inkie dinkie: A very small low power light source.

inking: In *computer graphics*, using a stylus to sketch freehand.

inky: See *inkie*.

inlay: A video effect where one video image is inserted into part of another.

in-line tube: A type of colour *cathode ray tube* with its three *electron guns* in a single plane rather than in a triangular formation as found in a *delta gun tube*. See **Figure I.1**.

Inmarsat: See *International Marine Satellite Organization*.

in phase: See *phase (1)*.

in-point: In video editing, the first *frame (2)* that is recorded at the beginning of a video edit or the starting frame of an audio edit.

input: The data or signals entering a *computer*. More generally, the signal applied to any electronic device or system.

input impedance: The *impedance* that a device presents to its source.

input looping: US term for *loop through input*.

input/output (I/O): How a *computer* receives input and generates output.

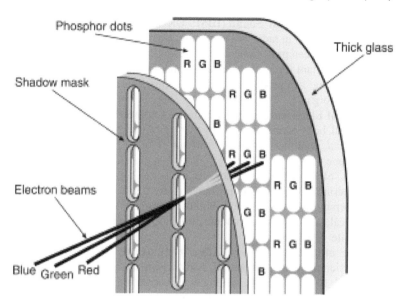

Figure I.1 In-line cathode ray tube

insert: **(1)** In video *editing,* any material replacing a section within an existing recording. **(2)** In film, it is usually a scene photographed separately and added during *editing,* such as a close-up shot of a letter or a newspaper. **(3)** In a *keyer,* the video source used to fill the hole cut by the *key signal* in the *background (1).* See also *fill.*

insert editing: *Editing* where new video material, audio or *time code* signals replace existing material, without introducing a visual disturbance at the edit point. A *videotape* editing system uses existing *control track* information which must already be on the tape (compare *assemble editing*).

Insertion Communication Equipment (ICE): BBC communication system using a line in the *field blanking period.*

insertion test signal (ITS): A video test waveform inserted in the *field blanking period* which can be used to assess the technical quality of the signal.

Institute of Amateur Cinematographers (IAC): An international organization based in the UK aiming to further interest and education in relation to all aspects of film and video making and associated visual arts. Its preferred title is IAC – The Film & Video Institute. (Internet: www.fvi.org.uk).

Institute of Broadcast Sound (IBS): UK institute founded in 1977 by sound balancers in radio and television, to provide a better interchange of ideas between practitioners in the various areas of broadcast audio. It is an important forum for all audio professionals working within broadcasting in production and *post production*. (Website: www.ibs.org.uk).

Institute of Electrical and Electronics Engineers, Inc. (IEEE): US technical body formed in 1963 by merging the *American Institute of Electrical Engineers* with the *Institute of Radio Engineers* . (Website: www.ieee.org).

Institute of Electrical Engineers (IEE): The largest professional engineering society in Europe, based in the UK, with a worldwide membership of over 130 000. It aims to promote the advancement of electrical, electronic and manufacturing science and engineering. (Website: www.iee.org.uk).

Institute of Practitioners in Advertising (IPA): Industry body and professional institute for the UK's advertising agency business, founded in 1917. It is a non-profit making organization funded by member subscriptions. (Website: www.ipa.co.uk).

Institute of Radio Engineers (IRE): An international society for scientists and engineers involved in the development of wireless communications, based in the USA, which merged with the *American Institute of Electrical Engineers* in 1963 to form the *IEEE*. (Hist.)

Institute of Videography (IOV): UK organization established in 1985 for raising the standards of video production within the professional video industry. Its membership includes video professionals from all areas of the moving image industry, including corporate production companies, *videographers*, broadcast professionals and those involved in new media technology. (Website: www.iov.co.uk).

Institut für Rundfunktechnik GmbH (IR T): German broadcast engineering research centre.

in-store video: A public video display for point-of-sale information, sometimes *interactive*.

instruction: In computers, a single command forming part of a *computer program*. A sequence of instructions defines the task for the computer to execute.

instructions per second (ips): A measure of how fast a computer can execute a *program*. It is usually quoted in *mips*, millions of *(machine code)* instructions per second.

insulation displacement connector (IDC): A form of connecting system where wires are terminated without removing their insulation. The connector cuts through the insulation to obtain a good electrical contact.

integrated circuit (IC): An electronic device in which many individual components are fabricated together in a single multi-pin package.

integrated receiver decoder (IRD): Alternative term for a *set top box*, sometimes applied to a professional rather than consumer product.

Integrated Services Digital Br oadcasting (ISDB): Japanese digital *broadcast* system developed by *NHK* to provide users with a combination of broadcasting and other services (e.g. conventional television and radio, *HDTV* and TV newspapers). It is based on *MPEG-2* and is very similar to the *DVB* system. Defined in standard ITU-R BO.1227-2. See **Appendix S**.

Integrated Services Digital Network (ISDN): An international standard enabling a digital *telecommunications network* to carry data *traffic* at the same time as voice *traffic* (telephone calls).

Integrated Services Television (ISTV): Japanese concept for a television set with a two-way communication function which can receive conventional analogue broadcasts, digital broadcasts, and other services. *ISDB* providers will also provide an index that classifies and identifies the programmes and information they of fer to allow users to select different services.

Intelsat: See *International Telecommunications Satellite Organization*.

intensification: The process of increasing the *density* of a previously developed photographic image.

intensity modulation (IM): Used in *optical fibre* systems to transmit a signal by directly altering the brightness of the light source.

interactive: A video game or *computer program* in which there is a two-way interplay between the *viewer (2)* or operator and the system.

interchange: (1) In television transmission, the ability to send television signals from one country to another using the same colour *encoding (1)* system, for example *PAL*. **(2)** In videotape recording, the ability to replay a recording made on a dif ferent machine working in the same tape *format (2)*.

intercom: A two-way voice communication *channel (1)* between locations where only one end can talk at any one time when a talk switch at their end is depressed.

intercutting: A method of *editing* in which varied shots of the same subject are put together in a sequence.

inter-dupe: A *duplicate (2)* colour negative film derived by printing from an *inter-positive*.

interface: (1) A circuit which converts the output of one system into a form suitable for connecting to another . **(2)** A means of communicating information between a computer system and its operator , e.g. *GUI* or *WIMP*.

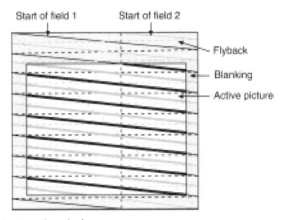

Figure I.2   Interlaced picture

interference: Disturbances to radio or television reception due to external sources of electromagnetic radiation.

inter-frame compression: Video data compression taking advantage of the *temporal redundancy* between *frames (2)*.

interlace: A video *scanning* system where the odd numbered lines in each *frame (2)* are scanned or drawn separately from the even numbered lines. The odd numbered lines form one *field*, the even lines make a second *field*, so one *frame (2)* comprises two *fields* (compare *progressive scanning*). See **Figure I.2** and **Appendix L**.

interlock: To interconnect two systems and hold them together physically or electrically.

intermediate: A general term for colour film master positives and *duplicate (2)* negatives printed on a film *stock* having integral colour masking *(2)*.

intermediate film (IF): An early television imaging system exposing 17.5mm film and developing it continuously in dedicated tanks, then scanning it for immediate transmission while still wet. It was used at the 1936 Olympic Games by the German television service.

intermediate level: A *dimmer* setting, usually pre-set, which lies between off and full up.

intermittent motion: The *transport* system in a film camera, *step printer* or *projector* in which each *frame (1)* in succession is moved into position and held stationary for the period of *exposure* or presentation.

intermodulation (IM): The generation of unwanted sum and dif ference frequencies when two or more signals pass through a non-linear system.

intermodulation product (IP): See *intermodulation*.

International Academy of Broadcasting (IAB): Academy based in Switzerland offering a post-graduate programme of studies dedicated to the arts and sciences of broadcasting. (Website: www.iab.ch).

International Alliance of Theatrical Stage Employees, Moving Picture Technicians, Artists and Allied Crafts of the United States, Its Territories and Canada (IATSE): US labour union representing technicians, artisans and craftspersons in the entertainment industry, including live theatre, film and television production. (Website: www.iatse.lm.com).

International Association of Broadcasting Manufacturers (IABM): Formed in 1976 to provide a UK forum for manufacturers of broadcast equipment or provision of services worldwide. It has a membership of over 120 companies. (Website: www.iabm.org.uk).

International Broadcasting Convention (IBC): Annual European broadcasting conference and exhibition held in September in Amsterdam, Holland. (Website: www.ibc.org).

International Electrotechnical Commission (IEC): Organization created in 1906 to prepare and publish international standards in the fields of electrical and electronic engineering. (Website: www.iec.ch).

International Marine Satellite Organization (INMARSAT): London-based organization set up in 1979 and privatized in April 1999 which operates a network of *satellites (2)* for transmissions of all types of international mobile services including maritime, aeronautical and land mobile. (Website: www.inmarsat.org).

International Radio and Television Society Foundation (IRTS): US foundation offering educational programmes on electronic media. (Website: www.irts.org).

International Recording Media Association (IRMA): An association representing the optical disc and magnetic tape industries. Previously known as the *International Tape Association*. (Website: www.recordingmedia.org).

International System of Units: See *Système International*.

International Tape Association (ITA): See *International Recording Media Association*. (Hist.)

International Telecommunications Satellite Organization (INTEL-SAT): *Satellite (2)* organization founded in 1964 and based in Washington, USA. Early Bird was their first *satellite (2)*. (Website: www.intelsat.int).

International Telecommunications Union (ITU): Located in Geneva, Switzerland and responsible for co-ordinating global *telecommunications* networks and services, including television and radio broadcasting. Founded in 1865 as the International Telegraph Union, it changed its name to the International Telecommunications Union in 1934. It is organized in sectors, including:

ITU-D: ITU-Development sector. It provides technical assistance to developing countries.

ITU-R: ITU-Radiocommunication sector. It establishes global agreements on the international use of radio frequencies, broadcasting and *satellite (2)* transmissions. It replaced the *CCIR* in 1994.

ITU-T: ITU-Telecommuncation sector. It establishes global agreements on *telecommunications* standards. It replaced the *CCITT* in 1994.

(Website: www.itu.int).

International Teleproduction Society (ITS): US association of members from the broadcast industry dedicated to promoting and furthering the use of video as a medium of communication.

International Television Association (ITVA): An association devoted to the business and art of visual communication, based in the USA. Its members work in video, film, distance learning, web design and creation, and all forms of interactive visual communications, along with all associated crafts. (Website: www.itva.org).

International Theatre Equipment Association (ITEA): (USA) In 1971, members of Theatre Equipment and supply Manufacturers Association and Theatre Equipment Dealers Association united to form TEA, Theatre Equipment Association. The idea behind TEA was to join forces into a single, permanent or ganization for the purpose of fostering and maintaining professional, business and social relationships among its members within all segments of the motion picture industry. In 1994, TEA became ITEA expanding to the international arena, including the countries of Canada, France, Italy , Singapore and Japan. Today, ITEA is representative of over 50 manufactures and 50 dealers across the country and around the world. (Website: www.itea.com).

International Visual Communications Association (IVCA): An affiliation of the *International Television Association* with the former *British Industrial and Scientific Film Association.* (Website: www.ivca.org).

inter-negative: (1) A duplicate colour negative film made from an *interpositive* and distributed to the production laboratories for making the *release prints.* **(2)** A duplicate colour negative film prepared directly from an original *reversal film* exposed in the film camera.

Internet: A global, decentralized communications network comprising millions of interconnected networks and their *computers*, invented by the US Department of Defense in 1973. See also *London Internet Exchange.*

Internet address: An *Internet Protocol address* specifically identifying a user on the *Internet* rather than one within a local *TCP/IP* network.

Internet Architecture Board (IAB): Originally set up in 1984 by the US Defense Department and the National Science Foundation as the Internet Activities Board, the IAB provides oversight of the architecture for the *protocols* and procedures used by the *Internet*. It is also responsible for editorial management and publication of the *Request for Comments document* series, and for administration of the various *Internet* assigned numbers. As part of the *Internet Society*, it also appoints the chair of the *Internet Engineering T ask Force (IETF)*. (Website: www.iab.org).

Internet Assigned Numbers Authority (IANA): The organization responsible for allocating *Internet* addresses worldwide. (W ebsite: www.iana.org).

Internet domain: Domain names as applied to the *Internet*. Certain conventions are used to define the type of user in the allocation of Internet domain names, including those listed in **Table I.2** and the country codes given in **Appendix P**. See also *domain name* and *Domain Name Service (DNS)*.

Table I.2　Internet domain name categories

Domain	Description
.ac	Academic institution (when country code used)
.co	Commercial organization (when country code used)
.com	Commercial organization (no country code used)
.edu	Educational establishment
.gov	Government organization
.info	Information services
.int	International organization
.mil	US military
.net	Network organization
,nom	Individual with personal site
.org	Non-commercial organization
.rec	Recreational or entertainment activity

Internet Engineering Task Force (IETF): The section of the *Internet Society* responsible for meeting the engineering requirements of the *Internet* by developing appropriate technical standards that are then approved by the *Internet Engineering Steering Group (IESG)*. (Website: www.ietf.org).

Internet Engineering Steering Group (IESG): The body that approves the standards produced by the *Internet Engineering Task Force (IETF)*.

Internet protocol (IP): The network layer *protocol* responsible for addressing and routing *packets* over the *Internet*. It does not guarantee the arrival of *packets* or the sequence in which they arrive at their destination.

Internet protocol address: An identifier for a *computer* or other device on a *TCP/IP* network. In *IP* version 4 it consists of a 32-bit numeric address written as four numbers between 0 and 255 separated by full stops.

Internet Research Steering Group (IRSG): A group which manages the Research Groups of the *Internet Research Task Force (IRTF)* and holds workshops focusing on research areas of importance to the evolution of the *Internet*. Its members include the IRTF chair, the chairs of the IRTF Research Group and other individuals from the research community .

Internet Research Task Force (IRTF): A group whose mission is to promote research of importance to the evolution of the future Internet by creating focused, long-term and small research groups working on topics related to Internet protocols, applications, architecture and technology . The IRTF chair is appointed by the *Internet Architecture Board*. (Website: www.irtf.org).

Internet service provider (ISP): A company providing access to the *Internet*.

Internet Society (IS): The organization that has administered the Internet since 1992. It functions through two main groups: the *Internet Architecture Board (IAB)* and the *Internet Engineering Task Force (IETF)*.

inter-positive: A colour master positive film printed directly from an original colour negative on to *intermediate stock*.

interpolation: (1) In digital video signal processing, generating missing picture detail by averaging information from the surrounding area within the picture *(spatial interpolation)* or in adjacent *frames (2) (temporal interpolation)*. **(2)** In digital audio signal processing, generating missing or corrupted audio *samples* by averaging information from the adjacent *samples*.

interpreter: A *computer program* which translates *source code* (written in a *high-level language*) line by line, to produce executable *object code* for subsequent execution.

interrupt: A method of gaining the immediate attention of a *computer*. The *computer* is programmed to respond in a particular way before resuming its previous task.

Intersputnik: Organization founded in 1971 to operate a global *satellite (2)*-based *telecommunications* system. Its 23 member nations are: The Republics of Afghanistan, Bulgaria, Belarus, Czech, Cuba, Geor gia, Germany, Hungary, Kazakhstan, Korea, Kyr gyzstan, Laos, Nicaragua, Poland, Syrian Arab, Tajikistan, Turkmenistan and Vietnam, Mongolia, the Ukraine, Romania, the Russian Federation and the Yemen.

intra-frame compression: Video data compression taking advantage of the *spatial redundancy* within a *frame (2)*.

inverse: In data presentation, the exchange of black and white to produce black lettering and lines on a white *background (4)* on a picture monitor and white on black in a printer .

inverter: **(1)** An *amplifier* which shifts the *phase* at its output by 180 degrees compared to its input. **(2)** A power supply that generates a high voltage from a low input voltage.

IP address: See *Internet Protocol address*.

IRE: Unit of video *amplitude* in 525-line television systems. The video signal from the bottom of the *sync pulses* to *peak white* is divided into 140 equal units. The video range from *blanking level* to *peak white* is 100 IRE units. Named after the *Institute of Radio Engineers* . See **Appendix K.**

iris: An adjustable *diaphragm (1)* to regulate the amount of light passing through it, e.g. in the *human eye* (see **Figure H.6**) or in a camera lens.

iris wipe: A *wipe* in the form of a circle. See **Figure W.1**.

I-signal: In the *NTSC* colour system, represents a *colour difference signal* on the orange-*cyan* axis (compare the *Q-signal*).

ISO: **(1)** The name ISO is derived from the Greek isos and is the worldwide name for the International Or ganization for Standardization, a non-governmental or ganization established in 1947 and based in Switzerland. (Website: www.iso.ch). **(2)** An independent video camera on a multi-camera production whose output is recorded separately .

isochromatic: Sensitive to colours in their corresponding relative visual intensity.

isolating transformer: A *transformer* with separate primary and second-ary windings to provide isolation and safety , usually from the mains supply.

isopropyl alcohol: An organic liquid solvent used for cleaning magnetic *heads (2)* and *tape paths*.

isotropic: Radiating uniformly in all directions.

isotropic antenna: A hypothetical omnidirectional point-source *aerial* which is an engineering reference for the measurement of *aerial* gain.

ITU-R BT.601: Formerly called CCIR Rec. 601, this ITU-R standard (see *International Telecommunications Union*) defines encoding parameters for digital *component video*. It defines the *sampling* systems, *matrix (2)* values and *filter (3)* characteristics for *YCbCr* and *RGB video* systems based on *luminance sampling* at 13.5 MHz and the *Cb* and *Cr* components at 6.75 MHz (See also *4:2:2*). The same frequencies are used for both 525-line and 625-line television systems.

ITU-R BT.656: Formerly called CCIR Rec. 656, this ITU-R standard (see *International Telecommunications Union*) defines the physical parallel and serial interconnection scheme for digital video signals coded to the *ITU-R BT.601* standard.

ITU-R BT.709-3: Part II of this recommendation describes the unique *HD-CIF* standard of 1080 lines by 1920 samples/line *interlace* and *progressively* scanned with an *aspect ratio (1)* of 16:9 at both 50 Hz and 60 Hz field and *frame rates* for *high definition television* programme production and exchange.

ITV Network Limited: The company that represents the interests of the UK *Independent Television* contractors. Up to 1998 it was called the *Independent Television Association*, or ITV Association. (Website: www.itv.co.uk).

J

jack: A cable-mounted plug of circular cross-section with two or three in-line contacts, usually used for *audio frequency* signals. See also *PO jack* and **Figure P.8**.

jackfield: A panel on an *equipment rack* containing rows of audio or video sockets for interconnecting equipment or signal paths. See also *patch panel, musa, U-link.*

jaggies: *Aliasing* around the edge of *computer graphics* characters or images. (Colloq.)

jam sync: Synchronizing a secondary *time code* generator to a master *time code* signal. It is used to ensure a clean *time code* signal is recorded during video *editing*.

Java: A general purpose *platform*-independent computer *programming* language developed by Sun Microsystems in widespread use to create *Internet applications*. (Website: www.java.sun.com).

jelly: A colour *filter (1)* or *diffuser* placed in front of a lighting source. See also *gel*. (Colloq.)

jib: The arm of a camera *crane (1)*.

jibbing: Moving the arm of a camera *crane* from one side to the other . Called a *tongue* in the USA.

jiggle: To add subtle film grain ef fects and motion to video generated material.

jitter: (1) Small irregular unsteadiness of the film in camera or *projector*. **(2)** The subjective ef fect of *flutter* in a reproduced video image. **(3)** Small irregular fluctuations in the timing of a signal. **(4)** Jerky movement due to faulty *animation*.

jog: A facility on film or video equipment to move backwards or forwards by a small amount.

join: A *splice* or edit between two shots.

Joint Electron Device Engineering Council (JEDEC): The original 1958 name of what is now The JEDEC Solid State Technology Association. It is the semiconductor engineering standardization body of the *Electronic Industries Alliance (EIA)*. (Website: www.jedec.org).

Joint Photographic Expert Group (JPEG): A committee established by *ISO* and ITU-T (see *International Telecommunications Union*) in 1982 to examine ways of reducing the data requirements for digital photographic images. Hence the name of the image data compression system they developed. Also called the Joint Picture Experts Group. (Website: www.jpeg.org).

joule (J): *SI unit* of energy, equal to the work done by a force of one *newton* when its point of application moves one metre in the direction of the force. Named after the nineteenth century English physicist James Prescott Joule.

joystick: A *computer* input device to give control in two or more dimensions. It is often used for positioning graphic elements or *wipe patterns* on a *vision mixer (1)*.

judder: A picture impairment caused by incorrect interpretation of motion in *standards converter* equipment. Objects or *background (3)* move across the picture with jerky rather than smooth motion.

jumbo: A wide roll of *magnetic tape* or photographic film produced by the coating machine before being *slit (2)* into strips of the final widths.

jumbo slide: A *transparency* of larger size than the conventional 24 × 36 mm *slide* (50 × 50 mm mount), usually 55 × 55 mm.

jump cut: In film and video *editing*, an interruption to the normal continuous action of a scene It produces a jarring visual join between shots which may or may not be intentional.

jumper: (1) A short piece of wire used to complete an electrical circuit. **(2)** A small plastic covered metal clip which fits over a pair of pins on a circuit board. When in place it completes an electrical circuit.

junior: A 2 kW studio spotlight. (Colloq.)

just a bunch of disks (JBOD): A collection of optical or magnetic *disks* for storing data. There is no protection of the data in the event of a *disk* failure (compare *RAID array*).

K

Ka-band: The *electromagnetic spectrum* between 18 and 31 GHz used for *satellite (2)* communications. It is part of the *K-band* region. See **Appendix N.**

Kansas city: An obsolete standard for recording *computer* data or *programs* on audio cassette tape. (Hist.)

K-band: The *electromagnetic spectrum* between 10.9 GHz and 36 GHz used for *satellite (2)* communications. See **Appendix N.**

Kell factor: A correction for a reduction in apparent vertical *resolution* in *interlaced* television systems compared to the number of lines. Commonly taken as about 0.7.

Kelvin (K): *SI unit* of thermodynamic temperature, equal to 1 degree Celsius, but on a scale where zero corresponds to − 273.16 degrees Celsius. This is the absolute zero of temperature and so the Kelvin unit used to be called degrees absolute. Named after the nineteenth century British physicist and inventor Sir William Thomson, Lord Kelvin.

kernel: The core component of a computer *operating system.*

kerning: In typography, controlling the horizontal distance between characters to improve appearance.

Kestral: A wheeled *camera mount* made by Vinten and consisting of a platform with a seat and *camera mount* on the end of a *boom* arm which can be raised from about 0.5 metres to 2 metres from the ground. (Trade name.)

key: A digital signal which allows *encrypted* television programmes to be descrambled by permitted users.

keyboard: A man-machine *interface (2)* where information is entered on buttons or keys, as used in musical instruments and *computers.*

key code: A *barcode* on the edge of motion picture film which allows film *edge numbers* to be electronically read and inserted into an edit list.

keyer: (1) A vision mixer effect system which allows one signal (the *key signal*) to cut a hole in the main picture (the *background (1)*) into which can be inserted a different picture (the *fill (1)*, *insert (3)* or foreground) the key and fill signals may be derived from the same source. Thus *captions* may be *keyed* into a picture (compare *chromakey*). **(2)** An electronic circuit which generates a control signal (based on a *key signal*) for a video multiplier whose output will be combined elements of its two inputs (normally *background (1)* and *fill (1)* video).

key frames: Graphics of the principal objects and poses generated electronically and stored in a *computer* for use in an *animation* sequence.

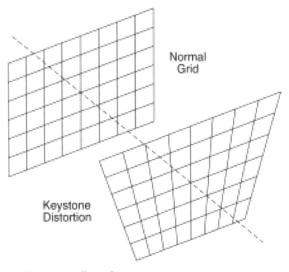

Figure K.1   Keystone distortion

key grip: The head *grip*.

keying: The process of selectively replacing parts of one video image with another.

Keykode: A machine readable *edge numbering* system. (Trade name.)

key light: The principal light illuminating the main subject in a scene. See also *modelling light*.

key numbers: See *edge numbers.*

key pad: A small keyboard or part of a keyboard, used for numeric entry . It has the numerals 0 –9, and may have +, – and other keys.

key signal: The video source in a *keyer* which shapes the hole cut in the *background (1)* image and has the complementary ef fect on the *foreground* signal. Sometimes called an *alpha channel*.

keystone distortion: Image distortion in an optical or video system causing both the vertical sides of a rectangle to conver ge. Called *trapezoidal distortion* in the USA (compare *trapezium distortion*). See **Figure K.1**.

kicker: A spot lamp used to give a highlight ef fect in side lighting.

kilo: A prefix denoting a numerical factor of one thousand, but often misused for *kilogram*. In computing, kilo is often taken as 1024, the nearest power of 2 to 1000, e.g. *kilobyte*.

kilobyte: 1024 *bytes.*

kilogram: *SI unit* of mass.

kilohertz (kHz): One thousand cycles per second. See *hertz*.

kilovolt ampere (kVA): A rating describing the total voltage applied and current flowing in a circuit. With unity *power factor* this will equal the power in watts. With a low *power factor*, (e.g. a fluorescent light source) kVA will be greater than the wattage, which implies a greater current flow than the wattage would indicate.

kilowatt (kW): 1000 *watts*.

kine: Abbreviation for *kinescope*. (Hist.)

Kinemacolor: The first commercial colour film process, patented in 1906 and developed by the Natural Colour Kinematographic Company, based in Hove, UK. The process was abandoned in 1915. (Hist.)

kinematographe: A film projector designed in 1896 by Oskar Messter , a German inventor. (Hist.)

kinematoscope: A device giving the illusion of movement by using still pictures on a spinning wheel, invented by Coleman Sellers, a nineteenth century American engineer. (Hist.)

Kineograph: See *flick book*.

kinescope: (1) US term for a film recording of television. Also called *kine*. See also *telerecording*. (Hist.) **(2)** US term for a *cathode ray tube* . (Hist.)

Kinetograph: An early electric film camera used in the 1890s to film scenes for the *Kinetoscope*, featuring such subjects as boxing matches, comic scenes and circus routines. (Hist.)

Kinetoscope: A peepshow viewing box designed in about 1893 by W. K.-L. Dickson, Thomas Alva Edison's assistant. (Hist.)

kludge: An untidy modification or adaptation, usually applied to *hardware*.

knee compression: A controlled non-linearity of the tonal reproduction in a video system, enabling it to handle a wider range of subject *contrast*.

Kodatrace: A matt-surfaced translucent plastic sheet material. (T rade mark.)

Ku-band: The *electromagnetic spectrum* between 10.9 GHz and 17 GHz used for *satellite (2)* communications. It is part of the *K-band* region. The band from 10.9 GHz to 11.7 GHz is used for *DBS* transmission. See **Appendix N**.

kukaloris: See *cookie (1)*.

kukie: See *cookie (1)*.

L

laboratory aim density (LAD): A value specified in the control of processing *intermediate* film.

lacing: To thread a motion picture *projector* or *magnetic tape* recorder.

lacquering: Coating the surface of motion picture film with a transparent layer to reduce or eliminate the visibility of scratches or abrasions.

lambert (L): Unit of *luminance (1)* (surface *brightness*) equal to one *lumen* per square centimetre. Named after the eighteenth century German physicist, mathematician and astronomer Johann Heinrich Lambert.

lamp-failure detector: A device to detect and report *projector* lamp failure. See *automatic lamp changer*.

lamp inertia: The time taken for the light output of a lamp to fall to zero when switched off.

land: On a *printed circuit board*, the exposed tinned copper pad onto which a component or one of its leads is soldered.

land line: A communication link using wires or cable, not radio.

landscape: A picture format where the image has greater width than height (compare *portrait*).

language: A set of representations, conventions and rules used to convey information. See also *assembly language*, *high level*, and *low level language*.

lantern: See *luminaire*.

lap dissolve: Abbreviation for *overlapping dissolve*.

laser: Light Amplification by the Stimulated Emission of Radiation. A device for generating a narrow , highly parallel beam of coherent *monochromatic light*, invented in 1960.

laserdisc: An optical video or data storage system read by a reflected *laser* beam.

laser printer: A high-quality printer in which a *laser* draws a fine pattern of dots on a photosensitive drum. The first laser printer was from IBM in 1975.

LaserVision: A *videodisc* system employing a *laser* beam to read analogue video and audio signals from a reflective disc. (T rade name.)

latching: A *thyristor* or *triac* being held in its conducting condition by the continued passing of current through the device.

latency: (1) The delay through a signal processing system. **(2)** The time taken for a *hard disk drive* to rotate half a revolution. This is the average time to locate a *sector* once the *head (2)* has arrived at a *track (3)* and is part of the overall average access time for a drive.

latensification: Intensification of an under-exposed image using controlled fogging by a light source before development.

latent image: The recorded invisible image contained in exposed but unprocessed photographic film.

latitude: The acceptable *exposure* range possible with a given photographic *stock*.

launcher: (**1**) US term for *feedhorn*. (**2**) A rocket used to place a *satellite (2)* into Earth *orbit*.

lavalier: A neck-hung *microphone*.

layback: In *post production*, transferring the finished audio track back to the master *videotape*.

layout: A pencil drawing showing the main features of the *background (4)* of a scene for *animation*.

lazy boy: See *pantograph*. (Colloq.)

L-band: (**1**) The *electromagnetic spectrum* between 500 MHz and 1600 MHz used for *satellite (2) downlinks*. Also refers to the specific band between 950 MHz and 1450 MHz used for mobile communications (see **Appendix N**). (**2**) *Wavelengths* between 1570 nm and 1610 nm used in *optical fibre* communications systems.

LCT arc: Low colour temperature arc lamp.

leader: (**1**) A length of film or *magnetic tape* at the start of a *reel* providing identification, alignment and timing information and protection. (**2**) A length of coloured non-magnetic tape attached to the beginning and end of some magnetic recording tapes.

leading: The vertical distance between rows of characters, in either typesetting or electronically generated text.

leading edge: The first rising or falling edge of a pulse (compare *trailing edge*).

least significant bit (LSB): The *bit* representing the smallest change in a digital word as it increments from 0 to 1, e.g. in a *byte*, changing the LSB from 0 to 1 adds 1 to the value of the byte (compare a most significant bit change adds 128).

left-hand circular polarization (LHCP): See *circular polarization*.

legs: See *tripod*. (Colloq.)

leisure time programming: Operating a *tape-slide* or multivision show at the operator's own pace, one *cue* at a time, rather than in *real time*.

lens: (**1**) An optical device of one or more *elements (3)* in an illuminating or image-forming system, such as a camera or *projector*. (**2**) The clear plastic cover of a pushbutton switch. (**3**) The part of the human eye which focuses an image onto the *retina*. See **Figure H.6**.

lens aperture: The effective opening of a lens system. It may be expressed as a fraction, *f-number* (the ratio of *focal length* to physical opening) or as a measured factor of transmission, *T-number*.

lens cap

16:9 letterbox

14:9 letterbox

Figure L.1   Letterbox presentation

lens cap: A light-proof and dust-tight cover to protect the *elements (3)* of a *lens* when not in use.

lens hood: An extension of the barrel of a *lens*, in solid or bellows form, used to restrict stray light from entering the *lens*.

lens mount: The part of the body of the lens housing which engages in the camera or projector to which it is attached. It may have a screw thread of standard size (e.g. *C-mount*) or a *bayonet mount* or other fixing.

lens turret: See *turret*.

lenticular screen: A projection *screen (1)* whose surface is covered with small lenses or ribs to improve image *brightness* by increasing the amount of light reflected in specific directions.

Leslie speaker: A *loudspeaker* mounted on a rotating baffle or surrounded by a rotating cylinder with a slot in it, often used with a keyboard musical instrument to give it a distinctive modulated quality. (Originally a trade name.)

letterbox presentation: Showing *widescreen* originated programme material with black bars at top and bottom on a narrower *aspect ratio (1)* display device, e.g. a 16 :9 *widescreen* programme on a 4 :3 television screen or a 2.35 :1 feature film on a 16 :9 television screen (compare *pillarbox*). See **Figure L.1**.

level: (1) The *amplitude* of an audio or video signal. **(2)** Sound volume, measured in *decibels*. **(3)** In *animation*, the position of a *cel* in a stack of cels, that nearest to the *background (4)* being level 1.

lever arm: In a *vision mixer*, the main control, usually with a T shaped handle, for changing between two selected sources using a *transition* or adding a *key*.

library: A store of picture and/or sound information available for reference and further use.

134

library material: Recordings intended for and held in a library .

library music: A music recording derived from a library , not an original recording made for the production concerned. It may be mood music specially written for use in audio and *audio-visual* productions.

library shot: A picture scene derived from a library , not an original recording made for the production concerned.

library slide: A *slide* derived from a library, not an original picture made for the production concerned.

lift: See *brightness*.

light: The section of the electromagnetic spectrum between *microwaves* and *X-rays* which includes part of the *infra-red* band, the *visible spectrum* and *ultraviolet light*.

light box: An illuminated panel against which film or *slides* may be examined. It is also used for tracing in *animation*.

light pen: A stylus with an optical sensor to allow interactive use with a *visual display unit*, as in early *computer graphics* systems. (Hist.)

light-emitting diode (LED): A semiconductor device which emits either visible light, when it is used for indicators and *alphanumeric* displays, or *infra-red* radiation, when it is used for position sensing and *remote control* applications.

lighting grid: The lighting gantry above a studio floor . See also *grid*.

lighting plot: A plan showing the number and position of *luminaires* in a television studio, theatre or set. Also the programme of lighting cues and states for their use.

lighting rig: The arrangement of *luminaires* and their accessories for a film or video production.

lightning display: A display on a *component video waveform monitor* showing *luminance* and *colour difference signals* to enable their levels and timing to be measured. See **Figure L.2.**

Lightworks: An electronic video *non-linear editing* system. (Trade name.)

lily: A coloured pattern imaged onto a film reel to act as a guide in processing and printing.

limes: *Front-of-house follow spots*.

limiter: A device restricting the maximum level of a signal.

limiting: Restricting the *amplitude* of a signal.

linear: Performance of a device or system where the output is exactly proportional to the input (compare *non-linear*).

linear distortion: *Distortion* which is independent of signal amplitude (within the operating limits of the system).

linear keying: *Keying* where the ratio of *foreground* to *background (1)* is determined on a linear scale by a control signal. This gives the best possible edge control and realism in the combined picture, especially using *anti-aliased* sources.

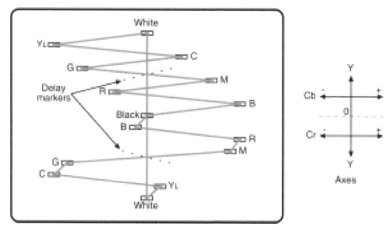

Figure L.2   Lightning display for 100% colour bars

linear movement: *Animation* giving movement at a constant speed, i.e. the same distance is moved between each *frame (2)*.

linear polarization: A *polarization* mode for an *electromagnetic wave* where the planes of the electric and magnetic components are in a fixed direction. Two linear polarization modes are used, horizontal and vertical, depending on the plane of the electric component, e.g. the wave in **Figure E.1** is vertically polarized (compare *circular polarization*).

line array: A single row of *charge coupled device* image sensors used in many imaging applications including *telecine*, *slide* scanners, flatbed scanners, printers, copiers and some facsimile machines.

line blanking: See *horizontal blanking period*.

line cord: US term for an *AC* mains power supply cable.

line crawl: A visual defect of images on *cathode ray tubes* with short-persistence phosphors, giving the impression that faint horizontal lines are moving up or down on the *screen (1)*.

line feed (LF): In word processing, a character which causes a printer or the display cursor to move to the next line. A *carriage return* character is needed to move to the start of the line.

line film: Ortho-sensitive photographic film processed to high contrast, for maximum separation between black and white.

line follower: In *computer graphics,* an input device which detects and traces lines in *vector* format.

line level: Audio signals at a nominal level of 0 *dBm*.

line-of-sight: A direct path with no obstructions between transmitter and receiver aerials which must be present in *radio frequency* communications or transmission systems using *VHF* or higher frequencies.

line period: The duration of a television *scan line* measured between the half-amplitude points on the leading edge of line sync pulses. Equal to exactly 64 microseconds in 625-line television systems. See **Figure C.5**.

line printer: A high-speed printer which prints a whole line at a time.

line segment: In *computer graphics*, a line bounded by two endpoints.

line scan: The horizontal sweep across a *cathode ray tube* picture or data display.

line sync: See *sync pulse (2)*.

line tests: Photographic tests using pencil drawings for smoothness of *animation*.

line-up slides: *Slides* in register mounts specifically produced for lining up the *projectors* of a multi-projection system.

line-up switch: A switch on *slide projectors* or on a *projector* control unit to turn on the lamps for test and alignment purposes.

lining up: (1) Setting up a camera in rehearsal ready for actual shooting. **(2)** Accurately adjusting equipment for optimum performance. **(3)** Accurately setting up two or more *slide projectors* for multivision.

lipstick camera: A small video camera, about the size of a lipstick holder, used in motor racing, attached to a skier , inside a cricket stump, for concealed filming or other specialist applications.

lip sync: Simultaneous picture and sound recording, to give exact correspondence between lip movement and spoken speech.

liquid crystal display (LCD): A display formed as a sandwich of two sheets of *polarizing* material with a liquid crystal solution between them. If an electric current is passed through the solution it aligns the crystals to either block or pass light to form the characters and graphics on the display.

liquid gate: In motion picture printing, an *aperture (2)* where the film is immersed at the time of *exposure* in a liquid of the same refractive index as the base. This avoids surface scratches on the negative being printed.

liquid printing: Also called *wet printing*. See *liquid gate*.

lith film: An extremely high contrast black and white photographic film.

lith mask: (1) A mask produced by photographing artwork on *lith film* for *sandwiching* with a *transparency* in a *slide mount*. **(2)** A high contrast mask used for producing a composite image in optical printing on to a single *slide*.

live: (1) Describing information processed as it occurs in *real-time,* as opposed to recorded. **(2)** Describing an audio monitoring environment with large *reverberations*.

luminaire: A complete lighting unit (i.e. lamp, lens, casing and reflector) in a studio or theatre giving some control of the colour , size, shape and sharpness of the light, e.g. *broad, flood* or *spot.* Also called a *lantern,* or a *fixture* in the USA.

luminance: (1) The *brightness* of a surface emitting or reflecting light. **(2)** The part of a video signal conveying the *brightness* of the image, given the letter designation Y. Also called *luma.*

luminosity: See *brightness.*

luminous flux: The rate of flow of visible light from a source, measured in *lumens.*

lunar outage: See *outage.*

lux (lx): *SI unit* of illumination (light falling on a surface), equivalent to the incidence of one *lumen* per square metre.

M

M II: A *videotape* recording system of *broadcast standard* developed by Panasonic (part of Matsushita). It uses 12.5mm ($\frac{1}{2}$ inch) *metal particle* tape for component recording of *luminance* and *colour difference signals* on separate tracks. (T rade name.) (Hist.)

machine code: The lowest level of *programming* for a *computer*. The individual *instructions* are in a numeric code, usually in *binary* or *hexadecimal* notation. See also *object code*.

machine room: See *apparatus room*.

M and D: In motion picture film, *masters* and *dupes*.

M and E: Music and effects. A sound track containing music and ef fects without speech or dialogue.

M&S mic: Middle and side *microphone*, a combination used for stereo sound recording.

macroblock: In *MPEG* video data compression, a group of four *luminance* sample blocks of 8 *pixels* by 8 lines each (totalling 16 *pixels* by 16 lines) and two *colour difference signal* blocks, used as the basis for *motion estimation* between *frames (2)*.

macro lens: A lens capable of very close-up focusing.

magazine: (1) A light-proof film container for use with a camera, printer or processing machine. **(2)** A compartmented *slide* carrier that holds *slides* ready for use in a *projector*.

magenta: The *complementary colour* to green.

magnetic cartridge: A *gramophone* pick-up based on a moving magnet assembly. (Hist.)

magnetic film: A strip of magnetically coated or striped material having *perforations* like those of photographic film.

magnetic flux: The concentration of magnetic lines of force in a region, determined by the field intensity and the *permeability* of the medium. It is measured in *webers*.

magnetic tape: A storage or recording medium in the form of a long narrow strip of magnetically coated plastic.

magnetometer: An instrument for measuring magnetic fields.

magneto optical disk (MOD): A high capacity *computer* storage device which uses a *laser* beam to heat up pits on the surface of a magnetic *disk* to the point where their magnetic state can be changed by a recording *head (2)*.

magnetron: A power *thermionic valve* used to generate *microwaves*.

magnification factor: See *Q-factor*.

mag-opt: A motion picture print with both magnetic and photographic (optical) sound tracks.

mail server: See *e-mail server.*

mains hum: *Hum* originated by *interference* from an *AC* mains power supply.

main title: The section of a motion picture film in which the name of the production and the principal *credits* are displayed.

maltese cross: Part of a mechanism providing intermittent frame by frame movement in a film *projector.* See *geneva movement* and **Figure G.2.**

Manchester code: A method of *channel coding* which removes any *DC* component, is *self clocking* and has a limited *bandwidth.* It is used in *Ethernet LANs.* It is not the same as *bi-phase mark* code.

Marata: A rigid material for rear-projection *screens (1),* of fine grain and dark colour to minimize the ef fect of extraneous light. (T rade name.)

margin: The amount of signal in *decibels* by which the *microwave link* or *satellite (2)* system exceeds the minimum levels required for operation.

mark: One of the two states in a binary data transmission system, usually representing a binary one (compare *space*).

marker: In a *computer graphics* display, a user-defined symbol which can be repeated as required.

mark in: To select the point where an edit will start.

mark it: A command at live action filming to have the *slate* put in so as to identify the *shot* about to be taken.

mark out: To select the point where an edit will end.

married print: A motion picture print with both picture and sound on the same strip of film correctly in *sync* for projection.

marrying-up: Preparing picture and sound negatives in the laboratory for making *married prints.*

maser: Microwave Amplification by Stimulated Emission of Radiation. A device for generating a narrow , highly parallel beam of coherent *microwave radiation.*

masking: **(1)** A *psychoacoustic* effect whereby a sound suppresses the subjective effect of *noise* in the adjacent frequency band. **(2)** In colour photography, a method of correcting unwanted absorption in the dye components by the use of complementary colour images, which may be integral to the material or separate from it. **(3)** A combination of a normal photographic record with a low *density* image of inverse tonal values to reduce its ef fective contrast in printing or projection. **(4)** The indication of the unwanted areas of an illustration, either by means of an opaque cut-out overlay, or by lines marked on a *transparency,* or on the back of an illustration. **(5)** Limitation of a projected picture by combination of the *aperture (2)* in the *projector* and black borders surrounding the *screen (1)*. **(6)** The ability to adjust the *colour balance* by matrixing the *RGB* signals. Used to match colour film *primary colours* to the television standard. **(7)** A *key* setup using a *wipe pattern*, adjustable box shape or external mask signal to prevent unwanted areas of a *key signal* from cutting a hole in the *background (1)*.

Masking pattern adapted Universal Sub-band Integrated Coding And Multiplexing (MUSICAM): A form of audio data compression using *psychoacoustic* coding. It uses 32 filter banks each 750 Hz wide and analyses the audio signal in 24 ms blocks. It is the same as *MPEG* Layer II coding and is used in *DAB* and *DVB*, and is the preferred option with European *DVDs*.

mast: A tower or pole on which an aerial can be mounted to increase its height above the surroundings.

master: (1) The original 16mm reversal film exposed in the camera, after processing. **(2)** A special positive print made from an original negative for protection or duplication rather than projection. **(3)** The final version of any programme (video, audio, or *tape-slide*) from which *show copies* will be made. **(4)** An output gain control on an audio mixing desk. **(5)** A similar control on a vision mixer . **(6)** A similar control on a lighting desk.

master antenna television (MATV): Term used for television distribution from a central *aerial* in flats and apartment blocks.

master control room (MCR): The technical operational area where all incoming and outgoing feeds for a television broadcast centre are co-ordinated for communications, recording or transmission purposes.

match dissolve: A *dissolve* where one object is seen in different settings but occupying the same position on the *screen (1)* throughout the *dissolve*.

matching transformer: A device for coupling two systems of differing *impedance,* e.g. matching a *microphone* to an *amplifier* input circuit.

match line: In *animation*, a line showing exactly where an object on an upper *level (3)* has to appear to pass behind one on a lower *level (3)*.

matrix: (1) An electronic circuit system used to route several audio or video signal sources by different paths to different destinations. It is internally organized as an array of columns and rows. Also called a *router (2)* or *routing matrix.* **(2)** An electronic circuit to convert signals between different formats, e.g. to convert *RGB* to *luminance* and two *colour difference signals.*

matt: Describes a surface which reflects light uniformly in all directions, e.g. a projection screen.

matte: (1) An opaque mask limiting the area of a picture which is exposed in special effects. It may be a cut-out *aperture (1)* or a high *density* image on film, while in video it may be electronically generated to blank off the appropriate part of the signal. See also *travelling matte*. **(2)** A *full-field* colour video signal which can be used to fill areas of keys and borders. It is adjustable in *hue, saturation* and *brightness* to allow any colour to be used.

matte box: A holder, usually rectangular in shape, fitted in front of the lens of a film or video camera. It acts as a lens-shade and can hold optical and other *filters (1)* and shaped *apertures (1)* for special effects.

matte fill: A *matte* video used to fill the *hole* in a *key* effect.

maximum use of surface area: See *musa*.

maxwell: Unit of magnetic flux, equal to the flux through an area of 1 cm^2 normal to a uniform induction of one *gauss*. Named after the nineteenth century Scottish physicist James Clerk Maxwell. The *weber* is the preferred *SI unit*.

mean time between failures (MTBF): A reliability assessment.

Mechanical Copyright Protection Society (MCPS): The UK organization that acts to protect the interest of writers/arrangers of music when their works are mechanically reproduced.

media access control (MAC): A *protocol* layer in data *network (1)* systems.

media server: See *video server*.

media transfer: Copying from one medium to another, e.g. from *tape-slide* to video.

medium close shot (MCS): Showing a subject from about the chest up.

medium close-up (MCU): Showing the head and shoulders of a subject.

medium Earth orbit (MEO): An *orbit* typically between 10 000 km and 20 000 km above the Earth. See **Figure O.4**.

medium frequency (MF): The region of the *electromagnetic spectrum* between 300 kHz and 3 MHz. See **Appendix N**.

medium long shot (MLS): Showing almost a full-length figure of a subject.

medium shot (MS): Generally showing a subject from the waist up.

megabyte: 1024 *kilobytes*, equal to 1 048 576 bytes.

megahertz (MHz): One million cycles per second. See *hertz*.

menu: A list of options presented on a display to the user of a system. For example, the list of picture control options available to an animator in a *computer graphics* package or the set up options for a video camera presented in its *viewfinder (1)*.

metadata: Data about data. For broadcast programme material this might include the title, duration, time and date, copyright details, location or type of programme. Metadata has become a vital part of storing digital video and audio material in large *archives* to enable it to be found again easily.

metal-halide lamp: A compact mercury arc with metal halide additions, enclosed in a quartz envelope and often protected by a hard glass shield.

metallized screen: A projection *screen (1)* whose surface has been treated with metallic particles to give a mainly *specular* reflection.

metal evaporated (ME): A type of magnetic recording tape where ultra-fine metal particles are vacuum deposited onto the *backing.*

metal oxide: A combination of a metal and oxygen. The term is used to describe a type of magnetic recording tape.

metal particle (MP): A type of magnetic recording tape where ultra-fine metal particles are carried in a *binder* which is coated as a thin layer onto the *backing.*

metal tape: Magnetic recording tape with ultra-fine metal particles carried in a thin *binder* layer (*metal particle* tape) or vacuum deposited (*metal evaporated* tape).

Methods for Optimisation and Subjective Assessment in Image Communications (MOSAIC): A project set up by the European Union to produce test picture sequences and define assessment methods for subjectively analysing compressed and processed digital video signals.

metre: *SI unit* of length.

metropolitan area network (MAN): The interconnection of *computers* and *peripheral* equipment over an extended area, e.g. a town or city (compare *LAN*).

M-format, M-wrap: The *tape path* used in the *VHS* system.

mho: The reciprocal of an *ohm*. It is a measure of conductivity, equal to the conductance of a body whose resistance is one *ohm*. See also *siemens.*

mic: See *microphone.*

micro: (1) A prefix denoting a factor of one millionth, 10^{-6}. **(2)** Abbreviation for *microprocessor* or microcomputer. (Colloq.)

microbreak: A very short supply interruption.

micro cassette: A miniature cassette similar to a *compact cassette* but about one third the size. Used for dictation, music or data recording.

microdrive (1) A miniature data cassette containing an endless tape. (Hist.) (Trade name.) **(2)** A miniature *hard disk drive* developed by IBM which can be used in a *PC-card* slot on a *personal computer.*

microfilm: Photographic reproduction of data and/or illustrations in a size too small to read without magnification.

micrometre (μm): One millionth of a metre.

micron: One millionth of a metre. *Micrometre* (μm) is the preferred term.

microphone: A *transducer* to convert sound to an analogue electrical signal.

microprocessor: (1) An integrated circuit chip containing a logic processor, memory, control elements and interface circuits. The first microprocessor, the Intel 4004, was developed in 1971 by a team led by an American, Marcian Hof f. (2) Loosely used as the name for a programmed microcomputer.

microprocessor unit (MPU): See *microprocessor.*

micro-reciprocal-degree: See *mired.*

microsecond (μs): One millionth of a second.

microwave: Electromagnetic waves used for radio transmissions between about 1 GHz and 300 GHz (corresponding to wavelengths of between 30 cm and 1 mm respectively). They are used for transmitting television pictures between dif ferent sites using *dish (1) aerials* and in *telecommunication satellite (2)* systems.

microwave link: A *point-to-point* video or *telecommunications* link established between two fixed or mobile locations using *line-of-sight* and *dish (1)* or *rod aerials* and *microwave* radio frequencies.

Microwave Multipoint Distribution System: See *Multichannel Multipoint Distribution System.*

middle fairing: A *fairing* used to blend *animation* movement from one speed to another.

midi: See *Musical Instrument Digital Interface.*

mike: See *microphone.*

million instructions per second (mips): See *instructions per second.*

millisecond (ms): One thousandth of a second.

miniature circuit breaker (MCB): A re-settable alternative to a fuse.

MiniDisc (MD): A system from Sony using *Adaptive Transform Acoustic Coding (ATRAC)* audio data compression to record up to 74 minutes of stereo digital audio on a 6cm diameter optical disc. (T rade name.)

mini mac: A compact quartz *luminaire* used as a floodlight.

minimum object distance (MOD): In a camera lens specification, the closest distance separating the object of interest and the front surface of the lens.

mips: million instructions per second: See *instructions per second.*

mired: MIcro-REciprocal-Degree. A unit of *colour temperature*, being one million divided by the *Kelvin* value. Thus *daylight* of 5500K corresponds to 182 mired. See **Table C.1.**

mix: (1) To blend audio or video sources together creatively . (2) A visual effect equivalent to a *dissolve.*

mix down: The process of mixing audio signals from a lar ge number of *tracks* to a smaller number , e.g. an 8-track master to a stereo pair of *tracks.*

mixed feed: A video signal sent to a studio video camera of the output of a second camera for aligning a shot involving both cameras.

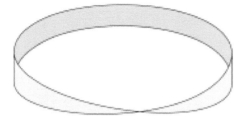

Figure M.1   Möbius loop

mix/effects (M/E): A subsystem of a *vision mixer (1)* where a combination of two or more images can be created. Each M/E typically includes two or three *buses (1)*, one or two *keyers* and a mixer.

mixing: The process of combining separate sound or visual sources to make a smooth combination or change.

mix-minus: US term for *clean feed.*

mnemonic: In computers, a short *alphanumeric* code which is an abbreviation for a *machine code* instruction. This makes generating and understanding the operation of *assembly language programs* easier.

mobile control room (MCR): An *outside broadcast* vehicle equipped and used as a *control room.*

Möbius loop: A length of tape or film joined to form a loop with a half-twist in it so that each surface will pass any point, used in some magnetic cartridges and printer ribbons. See **Figure M.1**.

model sheet: An animator's reference sheet showing the specifications and relative sizes of cartoon characters.

model shot: A film or video shot in which models are used instead of real objects.

modelling: The process of creating an *object* for use in a *computer graphic* or *animation* sequence.

modelling light: Light(s) placed to reveal the form and texture of the object being photographed. See also *key light.*

modem: MOdulator/DEModulator. A device used for sending digital information over audio circuits, such as telephone lines.

modified NTSC: A colour television system with *NTSC* coding but using a colour *subcarrier* frequency of 4.433 618 75 MHz (as used for *PAL*) instead of the normal *NTSC* value. This is sometimes more convenient for triple standard *video cassette recorders* and monitors.

modulate: To alter a characteristic (e.g. *amplitude*, frequency or phase) of a *carrier signal* at one frequency in correspondence with changes in a second signal at a lower frequency, e.g. to impress a signal on a *carrier wave.*

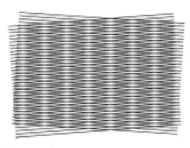

Figure M.2   Moiré patterning

modulation: Alteration of a characteristic (e.g. *amplitude*, frequency or phase) of a *carrier signal* at one frequency in correspondence with changes in a second signal at a lower frequency , i.e. changes one signal due to another signal. See also *amplitude modulation (AM)* and *frequency modulation (FM)*.

modulation depth: The amount of *modulation*, particularly applied to *amplitude modulation.*

modulation noise: *Noise* which appears only when *modulation* is present.

modulation transfer function (MTF): A measure of the performance of a lens system based on its ability to resolve detail at increasing frequencies. It is ef fectively a graph of *contrast* against *spatial frequency.*

modulator: (**1**) A device for impressing a signal on a higher frequency *carrier.* (**2**) A device for producing audio ef fects where one sound is *modulated* by another, e.g. a *ring modulator.*

module: (**1**) In *audio-visual* practice, a self-contained programme that may be used alone or be integrated as part of a lar ger programme. (**2**) In electronics, a *printed circuit board* or assembly containing electronic components which is part of a lar ger system.

moiré: Visual patterns formed by *interference* between two sets of regular divisions, such as the combination of a video *raster* with a striped object in the scene. It can also be produced by *CCD imagers* because of the grid arrangement of the *pixels.* See also *alias (1)* and **Figure M.2.**

Mole: Mole Richardson. A wheeled *camera mount* with the camera on a counterbalanced arm that can be raised or lowered. (T rade name.)

monitor: (**1**) A video display device not fitted with *radio frequency* receiving circuits. It is used to check the quality of what is being rehearsed or transmitted or to display text or graphical information. (**2**) To view television images, video *waveforms* or audio signals to check the quality of what is being rehearsed or transmitted. (**3**) A simple and permanently available first-level *computer operating system.* (Hist.)

monitor loudspeaker: A high-quality *loudspeaker* used to listen to the results of sound mixing and/or recording.

mono: Monaural or monophonic, a single audio channel.

monochromatic light: Light of a single *wavelength* (colour).

monochrome: Reproduction in a single colour , normally as a black and white picture.

monomode fibre: See *single mode fibre.*

monophone: A single earphone with a handle, often used at exhibitions.

monopod: A camera support with a single leg.

montage: A complex series of rapidly changing images and visual effects, often with several pictures being displayed simultaneously .

Moonshot: A lorry mounted hydraulic platform which can be raised to a height of about 46 metres.

Moore's law: In 1965 Gordon Moore, a co-founder of Intel, observed in a speech that each new computer chip contained twice the capacity of its predecessor and was released 18 to 24 months after . This doubling in computing power every 18 months is known as Moore 's law.

morphing: Short for metamorphosis. In *computer graphics*, changing one *object* into another over a series of intermediate images.

mosaic: (1) The pattern of *elements (4)* in a *CCD imager.* **(2)** A graphic effect where the picture, or part of the picture, is divided into small equal sized rectangles or squares. Each contains the average colour and brightness of the original picture area. It is often used to prevent a speaker's face from being recognized in an interview . **(3)** See *Methods for Optimisation and Subjective Assessment in Image Communications.*

most significant bit (MSB): The *bit* representing the largest change in a digital word as it increments from 0 to 1, e.g. in a *byte*, changing the MSB from 0 to 1 adds 128 to the value of the byte (compare a *least significant bit* change, which adds 1).

motion blur: A post production ef fect used to illustrate movement and speed.

motion estimation: In *MPEG* video data compression, determining a *motion vector* for a *macroblock* in an image to be coded from a previous or following *frame* so that the most efficient coding for the *macroblock* can be achieved.

Motion Picture Association of America (MPAA): US trade association founded in 1922 to put forward the views of motion picture producers and distributors, it now includes those involved in television, cable and home video programme production. Its services include advertising administration, copyright protection and guidance, market research, rating system guidance, technology evaluation and title registration. (Website: www.mpaa.org).

Motion Picture Editors Guild (USA): Formed as the Society of Motion Picture Film Editors in 1937, the Motion Picture Editors Guild became part of *IATSE* in 1944. (Website: www.editorsguild.com).

Motion Picture Export Association of America (MPEAA): The section of the *Motion Picture Association of America* providing internationally related services.

Motion Picture Experts Gr oup (MPEG): The original name of the *Moving Picture Experts Group.*

Motion Picture Sound Editors (MPSE): Founded in 1953 as a US organization for professional sound and music editors who work in motion pictures and television. (W ebsite: www.mpse.org/index.html).

motion vector: In *MPEG* video data compression, a horizontal and vertical offset for a *macroblock* from a previous or following *frame* to match as closely as possible a *macroblock* in the current *frame* being coded.

motorboating: Low-frequency *noise* which may be caused by instability in a sound *amplifier,* or by the influence of *perforations* on the sound track of 35mm motion picture film.

mount: (1) A carrier for a still film *frame (1)* or *transparency.* See *offset mount* and *register mount.* (2) See *camera mount.* (3) The assembly an *SNG dish (2) aerial* is fitted onto, including all necessary mechanical controls.

mouse: In *computer graphics,* a small hand-held device which can be moved freely over a horizontal surface. Its movements are conveyed to the computer to precisely control the position of a pointer on the display screen. It was invented by Douglas Engelbart of the Stanford Research Center in the USA in 1963.

moving coil: An electromagnetic *transducer* consisting of a coil of wire which can move in a magnetic field. It is used in electrical meters, some *microphones* and *loudspeakers.*

moving iron: A type of alternating current electrical meter consisting of a soft iron armature moving within a coil of wire.

Moving Picture Experts Group (MPEG): A working group of *ISO/IEC* established in 1988 and in char ge of the development of standards for coded representation of digital audio and video. They developed the *MPEG-1, MPEG-2* and *MPEG-4* standards, continuing with *MPEG-7.* (Website: www.cselt.it/mpeg).

Moviola: A film *editing* machine, allowing picture and sound films to be run in *synchronism.* (Trade name.)

MPEG-1: The video and audio data compression standard on which products such as Video CD and MP3 are based. Its title is 'Coding of moving pictures and associated audio for digital storage media at up to about 1.5 Mbit/s' and it is in five parts. See **Appendix X.**

MPEG-2: A set of video and audio data compression standards defined by the *Moving Picture Experts Gr oup* on which systems such as digital broadcasting and *DVD* are based. It also includes a system level specification on how compressed data streams can be combined and *multiplexed* to form services for transmission or storage and retrieval applications. The standard is in nine parts. See **Appendix X**.

MPEG-4: An object-based compression standard finalized in 1999 which provides the technological elements to integrate the production, distribution and content access rules for digital television, interactive graphics and interactive multimedia, e.g. *World Wide Web*. See **Appendix X**.

MPEG-7: A standard called 'Multimedia Content Description Interface ' describing the multimedia content data to interpret the information 's meaning. This can be passed to, or be accessed by , a device or a computer program. The elements that *MPEG-7* standardizes support a range of applications.

Multi-channel Audio Digital Interface (MADI): An interconnection standard carrying up to 56 channels of *AES/EBU* digital audio on 75-ohm *coaxial cable* and *BNC* connectors. It runs at a data rate of 125 Mbit/s. Sometimes known as Multi-channel Asynchronous Digital Interface.

Multichannel Multipoint Distribution System (MMDS): A *pay TV* system using *microwave links* rather than cable. Also called *Microwave Multipoint Distribution System* or *wireless cable*.

Multicore: (1) A cable with a number of separately insulated conductors within one sheath. **(2)** Solder wire that has several cores of flux running along its length. (T rade name.)

multi-frequency control: A control system where dif ferent channels are controlled simultaneously by several signals on one tape track using different frequencies, sometimes coded in pairs.

multi-image: US term for a presentation with several *projectors* on any *screen (1)* area.

multimedia: Originally a presentation employing a mixture of media, e.g. *tape-slide*, motion picture, video and live action, but now broadened to include almost any communications using any mixture from a variety of different media, often for entertainment. It can also include interactivity.

multimode fibre: *Optical fibre* with a *core (2)* diameter of typically 50 μm or 62.5 μm in which more than one mode of light propagation takes place (compare *single mode fibre*).

multiplane animation: Cartoon *animation* with *cels* placed on several planes at different distances below the camera.

multiple channel per carrier (MCPC): Using *time division multiplexing* to transmit several services on a single *satellite (2) transponder.* The system may include *statistical multiplexing* to provide improved and consistent picture quality.

Multiple Sub-Nyquist Sampling Encoding (MUSE): A Japanese bandwidth compression system to accommodate *high definition television* transmission within an existing analogue *satellite (2)* channel.

multiple system operator (MSO): An organization operating a number of *master antenna television* systems.

Multiple Unit Steerable Antenna (musa): An *aerial* consisting of an *array (1)* of *elements (2)* producing a beam whose direction can be varied by adjusting the *phase* relationships between the *elements (2).*

multiplex: (1) A time-sharing process where several dif ferent signals are transmitted over a single signal path. **(2)** In *MPEG-2*, a combination of the video, audio and data components of several *services.* **(3)** In the UK's digital terrestrial television system, the digital content of one of the six allocated *UHF* frequency channels. **(4)** A multi-screen cinema.

multiplexed analogue component (MAC): Coding an analogue component colour television signal using time division and compression. It was the European standard intended for *direct broadcast by satellite* transmission and was capable of extension to higher *resolution.* There were several variants using letter prefixes – see *B-, C-, D-* and *D2-multiplexed analogue component* . The system was overtaken by the rapid adoption of digital broadcasting technologies. (Hist.)

multiplexer: (1) A device to enable images from several sources (*slides* or motion picture) to be transferred on to one medium such as video or film. **(2)** Part of a multi-channel digital television broadcasting system that combines several services onto a single data path. **(3)** A device to combine several electrical or optical signals.

multipoint conference unit: A device that links three or more locations for interactive *video conferencing.*

Multipoint Distribution System (MDS): See *Multichannel Multipoint Distribution System (MMDS)*.

multiscan: A picture monitor which can operate at several *scanning* rates.

multi-screen: An *audio-visual* system having a number of image areas presented simultaneously to the viewer .

multi-tasking: A *computer* system able to give the illusion of executing more than one task at a time.

multi-track: An audio *magnetic tape* recorder having four or more parallel tracks. Commonly 4, 8, 16 or 24 tracks.

multi-tracking: (1) A technique of sound recording with a separate track for each source to permit subsequent mixing and blending. **(2)** Building up an audio track by the addition of several successive stages.

multi-user: A *computer* system capable of interacting with more than one user at a time.

multivision: A general term for *multi-screen* or multi-image presentation.

Mu-metal: An iron alloy used as a magnetic *screen (2)* material, particularly in audio magnetic tape *heads (2)*. (Trade name.)

Munsell system: A method of defining colours using numbers to represent *hue, brightness* and *chroma,* developed by and named after Albert H. Munsell, an American painter.

musa: (1) *Maximum Use of Surface Area.* A type of video connector, used particularly on video *jackfields.* **(2)** See *Multiple Unit Steerable Antenna.*

Musical Instrument Digital Interface (MIDI): A system for transferring control information and data between suitably equipped musical instruments and other musical equipment, including computers. It does not carry audio, so MIDI files store note instructions and not sound data.

Musicam: See *Masking pattern adapted Universal Sub-band Integrated Coding And Multiplexing.*

music chart: A breakdown, frame by frame, of a music track so that an animator can work exactly to the beat.

music-track: An audio track that carries only music.

mute: (1) A picture record only , without associated sound track on the same film or tape. **(2)** To disconnect or suppress the sound.

Mutoscope: A mechanical *viewer (1)* better known in Britain as a 'What the Butler Saw ' machine and using the principle of the *flick book.* (Hist.)

Mylar: Polyethelene terephthalate, often used as a *base* material for *magnetic tape.* (Du Pont Chemical Corporation trade name.)

N

NAB cartridge: *Broadcast standard* endless loop *magnetic tape cartridge* standardized by the *National Association of Broadcasters* and made in three tape capacity sizes.

NAB spool: A *magnetic tape spool* with a large central hole standardized by the *National Association of Broadcasters* and available in a range of outer diameters.

nadgers: a small movement of a camera or *prop*. (Colloq.)

nanosecond (ns): One thousandth of a *microsecond*; 10^{-9} seconds.

narration: See *voice over*.

National Association of Broadcasters (NAB): US broadcasting industry trade association based in Washington whose name is now applied to standards which they have specified and to an annual convention and exhibition held in April at Las Vegas, USA. (Website: www.nab.org).

National Association of Radio and Television Broadcasters (NARTB): Original name of the American *National Association of Broadcasters (NAB)*. (Hist.)

National Association of Theatre Owners (NATO): The largest exhibition trade organization in the world, representing more than 19 000 movie screens in all 50 US states and in more than 20 countries worldwide. Its headquarters are in North Hollywood, California. (Website: www.hollywood.com/nato).

National Bureau of Standards (NBS): Former name of the US *National Institute of Standards and Technology*. (Hist.)

National Cable Television Association (NCTA): US cable industry trade association. Founded in 1952, it aims to provide a single, unified voice on issues affecting the US cable industry . It also hosts the cable industry's annual trade show. (Website: www.ncta.com/home).

National Institute of Standards and Technology (NIST): The central measurement laboratory of the US federal government, established in 1901 and called the *National Bureau of Standards* until 1989.

National Museum of Photography , Film & Television (NMPFT): A museum founded in 1983 in Bradford as part of the UK' s National Museum of Science and Industry . (Website: www.nmpft.org.uk/home.asp).

National Physical Laboratory (NPL): The UK's national measurement standards laboratory. (Website: www.npl.co.uk).

National Television Systems Committee (NTSC): The committee set up in 1941 which defined an analogue colour television system for America, first demonstrated in October 1953. Hence the name of the standard they defined (strictly it should be called NTSC-II to distinguish it from the *monochrome* NTSC-I system). NTSC has 525 lines per *frame (2)*, 59.94 *fields* (29.97 *frames (2)*) per second and a 3.579 545 MHz *subcarrier*. It is defined in standard SMPTE 170M (see **Appendix U**) and used mainly in North America, Japan and parts of South America. (See **Appendix I** for a complete list.)

National Transcommunications Limited (NTL): UK company providing transmission and broadcast services for the ITV network, Channel 4, S4C, Channel 5 and most UK independent radio stations. It is also the largest cable group in the UK. (W ebsite: www.ntl.com).

Near Instantaneously Companded Audio Multiplex (NICAM): A digital coding system used in stereo sound transmission for television in many European countries. The sound signals are digitally processed to reduce their *dynamic range* for transmission, then expanded in the *receiver* to recover the original range of the signals.

near letterpress quality (NLQ): Describes the performance of *alphanumeric* printer systems.

negative: A developed photographic image in which the tones (and colours where they appear) are the reverse of those in the original scene (compare *positive*).

negative feedback: A condition where some of the output of a system is connected back to the input in *anti-phase*, frequently used in electronic *amplifiers* and control circuits.

negative temperature coefficient (NTC): A substance or component whose electrical resistance falls with rising temperature.

neg-pos: Photographic system in which the film exposed in the camera is processed as a negative image and used to print a separate positive.

neons: A term in *rostrum camera* work for neon-like lettering ef fects.

net: A logical linking of fixed points *(pins)* in a *computer aided design* display.

network: (1) A collection of *nodes (2)* interconnected by transmission paths. See also *local* and *metropolitan area networks*. **(2)** An affiliation of television or radio stations, particularly in the USA.

Network File System (NFS): A protocol developed by Sun Microsystems and Netware which allows groups of computers to access each other 's files as if they were locally stored.

network information table (NIT): *DVB* mandatory data table which is transmitted in the *DVB* data stream. It carries details of the physical organization of the *multiplexes (2)* and *transport streams* carried on a given network. See **Appendix O**.

neutral colours: The range of grey levels from black to white which have no colour. In a colour television system, the *RGB* signals are the same size and *colour difference signals* are zero.

neutral density: Having approximately equal values of *density* to the whole *visible spectrum*.

neutral density (ND): A filter used to reduce light level without affecting colour.

newton (N): *SI unit* of force. The force needed to accelerate a mass of 1 kilogram at 1 m s^{-2}. Named after the seventeenth century English scientist and philosopher Sir Isaac Newton.

Newton's rings: Optical *interference* patterns caused when two surfaces separated by a very small distance are illuminated; for example, the film surface of a *slide* and its cover glass. Named after the seventeenth century English scientist and philosopher Sir Isaac Newton.

nibble: Four *binary bits*. Two nibbles make a *byte*!

NiCad: A heavy-duty rechargeable battery based on nickel and cadmium. (Trade name.)

Nicam: See *Near Instantaneously Companded Audio Multiplex*.

Nike: A wheeled battery powered *camera mount* with a long *boom* arm made by Chapman in the USA. (Trade name.)

Nipkow disc: A perforated mechanical television *scanning* wheel. It was invented in the late nineteenth century by the German engineer Paul Nipkow. (Hist.)

nit: Unit of *luminance (1)*, equivalent to one *candela* per square metre. The word is French and derives from the Latin 'nitere', to shine. (Hist.)

nitrate: Abbreviation for cellulose nitrate, a highly flammable material once used as a film *base*. Once it starts to decompose, there is a serious risk of spontaneous combustion.

No-break: A power supply system which will maintain power without interruption in the event of a mains supply failure.

nodal head: A mount for a motion picture camera providing rotation and tilt centred exactly on the *nodal point* of the *objective lens*.

nodal point: The effective optical centre of an *objective lens* through which all light rays may be regarded as passing.

noddies: In an interview, *reverse shots* of the interviewer used as *cutaways* to bridge *jump cuts*.

node: (1) A point at rest in a vibrating system. (2) A networking term for a *station* on a *network*.

no good (NG): A common notation for a filmed *take (1)* or recording to indicate it is not to be printed.

noise: Unwanted sound or signals in a communication system, especially random electrical energy interfering with audio and video signals.

noise coring: See *noise slice*.

noise filter: Commonly an audio *low-pass filter* for suppressing *noise* outside the desired audio *bandwidth*, but may be a *band-pass filter* or a *high-pass filter*.

noise floor: The inherent residual *noise* level in a system.

noise gate: An automatic switch which *mutes (2)* an *audio frequency* path when the signal falls below a pre-set threshold level.

noise reduction: Using a circuit or system to reduce the subjective ef fect of *noise* on sound or picture quality .

noise slice: A method of reducing the perceived *noise* in a video signal by removing the information in its video *waveform* between predetermined levels, e.g. making anything below 5 per cent of signal *amplitude* at *black level (black slice)* . It is also called *noise coring* or just *coring.*

non-composite video signal: A video signal that contains picture and *blanking* information but not *synchronizing* signals.

non-flam: An alternative term for safety film.

non-linear: (1) Performance of a device or system where the output is not proportional to the input. **(2)** See *non-linear editing.*

non-linear distortion: *Distortion* which varies with signal amplitude.

non-linear editing: A computer-based editing system allowing *random access* to material to be edited.

non-return to zero (NRZ): A serial data stream in which the logic level remains at 1 during the transmission of consecutive data 1s (compare *return to zero*).

non-volatile memory: A memory system which retains data when its power supply is removed. See also *read only memory.*

nook light: A small light used in a confined space.

normally closed (NC): A relay or switch contact which opens in the operated condition.

normally open (NO): A relay or switch contact which closes in the operated condition.

North American Broadcasters Association (NABA): Part of the *WBU*, it provides a unified voice for its members on international broadcasting, journalistic, *telecommunications* and technical bodies. It also hosts and participates in relevant international conferences. (W ebsite: www.nabanet.com).

North American Broadcast Teletext (NABT): The *teletext* system used in North America, South America and the Far East, described in an open standard, the North American Basic Teletext Specification (NABTS).

north/south/east/west convention: In *animation*, the convention referring to the top of the field of a scene as North, the right hand edge as East, etc.

notch: A shallow cut-out made on the edge of a strip of motion picture film to actuate the light change in a printer, or to stop the film in certain forms of automatic presentation.

notch filter: A *filter (3)* which rejects a narrow range of frequencies. In audio, one is commonly used in the measurement of *total harmonic distortion.*

null: (1) To adjust to the lowest value or the exact centre value, e.g. setting an electronic circuit for least output with zero input. **(2)** In computing, a *string* space reserved in memory but containing nothing, i.e. of zero length.

number board: See *slate*.

Nyquist frequency: Named after Harry Nyquist and concerned with *sampling,* this is taken to be the minimum *sampling rate* for correct digital reproduction of an *analogue signal*. It is a *sampling rate* of twice the highest signal frequency to be reproduced. In practice, the ability to reproduce a signal sampled at exactly twice its frequency depends on the timing relationship between the *sampling* and the signal cycles. If the *sampling* pulses occur at the zero crossings, no output will result. Practical *sampling rates* are therefore usually well above this minimum frequency.

O

object: In *computer graphics,* a series of geometric surfaces that are grouped together to be *animated.*

object code: The result of translating a *source code computer program* into *machine code* using a *compiler* or *interpreter.*

objective lens: The main lens system of an instrument, such as a camera, producing an image of the object presented.

OB truck: *Outside broadcast* vehicle.

octal: **(1)** Numeric representation to the base 8, rather than decimal base 10. It used to be used in computing. **(2)** An eight-pin circular polarized connector used for *thermionic valves.* (Hist.)

octopus cable: A cable terminating in multiple plugs.

octave: A change in pitch representing a doubling of frequency .

oersted (Oe): Unit of magnetic field strength, equivalent to one *maxwell* per square centimetre. Named after the nineteenth century Danish physicist H. C. Oersted. The *tesla* is the preferred *SI unit.*

off-air: **(1)** Reception of a *broadcast* signal, often for quality monitoring purposes. **(2)** No longer transmitting.

off-line: **(1)** Isolated, not connected to the line, central *computer* system, etc. **(2)** Referring to *off-line editing.*

off-line editing: A method of video *editing* using low-cost equipment to produce a *rough cut* and/or *edit decision list (EDL)* prior to using expensive *broadcast standard* equipment for the final work.

off-scale: **(1)** Beyond the range of an instrument. **(2)** In motion picture printing, outside the range of the normal light point scale of a printer .

offset: A positive or negative time displacement in systems using *time code synchronization.*

offset feed: Type of microwave aerial in the form of a *parabolic reflector* with the *primary radiator* mounted to one side to avoid obscuring part of the beam (compare *front feed*). Used in most consumer *satellite* receiver systems. See **Figure O.1**.

offset lenses: See *perspective control lens.*

offset mount: A *slide mount* in which the *aperture (4)* is not in the centre of the mount. It is used to avoid keystone distortion when two or three *projectors* share a common *screen (1)* area.

ohm: *SI unit* of electrical resistance. Defined as the electrical resistance between two points on a conductor when a constant *electromotive force* of 1 *volt* produces an electrical current of 1 *ampere* in the conductor. Named after the nineteenth century German physicist Geor g Simon Ohm.

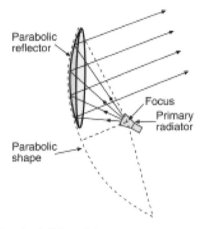

Figure O.1   Offset feed dish aerial

omega wrap: The *wrap* around the *head drum* of a *helical scan magnetic tape* recorder giving just less than 360 degrees of contact. See **Figure O.2** (compare *U-wrap*).

omnidirectional microphone: A *microphone* which is equally sensitive in all directions. Its *polar diagram* is therefore a circle (compare *directional microphone*).

OmniMax: A development of the *IMAX* system using the same size of film but presenting an even wider angle of view on a spherically curved *screen (1)*. (Trade name.)

on-air: Currently transmitting.

on-line: (1) 'Live', actively linked to the line, system, etc. **(2)** The final programme *editing* process in video *editing*.

Figure O.2   Omega wrap

one light: A motion picture print made at a single level of *exposure*, without alteration from scene to scene.

one-to-one position: A camera setting where the object and the image are the same size.

on screen display (OSD): The means by which adjustments to image size, geometry, colour and brightness are made on some computer displays.

opaquing: Painting animation cels after they have been traced.

open GOP: An *MPEG-2 Group Of Pictures* where the last frame of one GOP uses the first frame of the next (compare *closed GOP*).

open loop control: A control system not self-corrected by any form of *feedback* mechanism or circuit.

open loop gain: The *gain* of a *closed loop* system measured with the *feedback* path opened.

Open Media Framework Inter change (OMFI): A file format which provides open digital media interchange among applications and across different *platforms*. It represents how media is or ganized and lets users transfer source elements (video, audio, graphics and animation), ef fects and comprehensive *edit decision lists* from one application to another, in *sync*. (Website: www.avid.com/3rdparty/omfi/index.html).

open reel: A tape *transport* system with separate feed and take-up *spools*, as distinct from the enclosed path of a *cassette* or *cartridge*.

Open System Inter connect (OSI): A network model developed by *ISO* and published as ISO 7498 which represents the operation of network interconnections as a 7-layer *protocol stack*. Each layer performs a distinct function to encapsulate, protect or address the data to be transferred.

operating level: An audio signal level of +4 dBm, equivalent to zero VU. See *VU meter*.

operating system: A *computer program* or set of *programs* which provides the basic functions, user *interface (2)* and hardware *interfaces (1)* needed to use and operate a specific *computer*.

optical axis: The line of symmetry of the components of an optical system. For example, in a *zoom lens* or *compound lens* it corresponds to a line through the centres of the front and rear lens *elements (3)*. See **Figure P.10**.

optical character recognition (OCR): In a *computer* system, the capture of *characters* as data rather than graphics, to permit *editing* and further word processing.

optical fibre: A fine flexible glass or plastic filament used for the cable distribution of pulse-coded light signals in *telecommunications*, data networks and television. See also *core (2)*, *cladding* and **Figure O.3**.

optical printing: A method of photographic printing in which an image of the original is formed on the print *stock* by a copy lens.

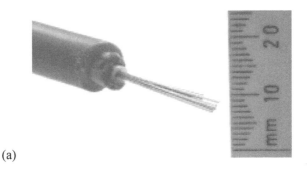

(a)

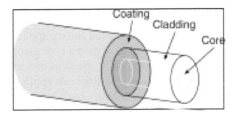

(b)

Figure O.3   Optical fibre construction

optical ROM: Laser disc technology for very high capacity storage of read-only data.

opticals: Modifications to the picture image made in the laboratory after filming is completed, such as *dissolves*, *wipes*, etc.

optical sound: See the preferred term *photographic sound*.

optical time domain reflectometer (OTDR): An electronic device which measures losses and distortion in an *optical fibre* system.

optical transfer function (OTF): See *modulation transfer function*.

optic nerve: The bundle of nerve fibres leaving the back of a *human eye* which carry electrical signals from the *retina* to the brain. See **Figure H.6**.

Orange Book: The standard for *Photo CD*.

orbit: The path of a *satellite (2)* around the Earth. It can take many forms, including *LEO*, *MEO* and *GEO*. See **Figure O.4**.

Orbital Test Satellite (OTS): UK *satellite (2)* programme in the 1970s. (Hist.)

Organisation Internationale de Radiodiffusion et Télévision (OIRT): The International Radio and Television Organization based in Prague which represented many countries in Eastern Europe until it merged with the *EBU*. (Hist.)

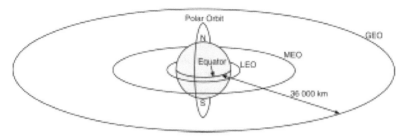

Figure O.4   Satellite orbits (drawn to scale)

orientation: (1) The alignment of the long axis of magnetic particles in *magnetic tape* manufacture to obtain maximum sensitivity. The direction of alignment differs according to the tape usage for audio, video or data recording. **(2)** The film position in 16 mm projection, emulsion-to-light (Type A) or emulsion-to-lens (Type B). **Figure O.5.**

original: (1) The film exposed in the camera, after processing. **(2)** The first video recording prior to copying or *editing*. **(3)** A final edited master.

original equipment manufacturer (OEM): Any manufacturer who sells its products to a reseller , e.g. Intel are an OEM who supply IBM with *microprocessor chips* for their *PCs*.

ortho: Orthochromatic, sensitive to only the blue and green regions of the *visible spectrum*.

Orthogonal Frequency Division Multiplex (OFDM): A digital *modulation* technique using hundreds or even thousands of closely spaced *carriers* to convey the digital information. A low data rate on each *carrier* accumulates to a high overall data rate. In *DVB*, either 1705 *carriers* (called 2k, and used in the UK) or 6817 *carriers* (called 8k) can be used.

orthogonal sampling: *Sampling* a video signal so that samples in each line are in the same relative horizontal position. The overall *sampling* pattern is therefore like a grid.

oscilloscope: An instrument used to make voltage changes visible over time. The word oscilloscope comes from 'oscillate', since oscilloscopes are often used to measure oscillating voltages.

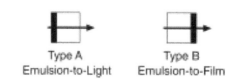

Type A   Type B
Emulsion-to-Light   Emulsion-to-Film

Figure O.5   16 mm film orientation symbols for projection

out-of-sync: When a picture and its associated sound are not in *synchronization*.

outside broadcast (OB): A broadcast originating from outside a studio centre. It is sometimes called a *remote*, particularly in the USA.

outage: Time when a *satellite (2)* or *microwave* communication link is not available, due to atmospheric conditions such as hail, snow or heavy rain, or, in the case of *satellites (2)*, periods when a *satellite (2)* is directly between the sun or moon and the Earth (called *solar outage* and *lunar outage* respectively).

out of vision (OOV): Commentary from a person who is heard but not seen.

out-point: In video editing, the first *frame (2)* that is not recorded at the end of a video or audio edit.

output impedance: The *impedance* that a device presents to its *load (2)*.

out-takes: Alternative *takes (1)* of a *shot* that are not used in the final *editing*.

overcranking: Filming at a higher speed than the intended projection speed, to slow down the action.

over-damped: *Damping* in which the response to a change is very slow. See **Figure D.1**.

overdub: The addition of another musical part to an already recorded performance.

overhead projector: A presentation device with a large horizontal projection *aperture (1)*. A lens and mirror/prism arrangement enable the image to be projected above and behind the user's head.

overlapping: In film or video *editing*, extending the sound track into the next scene for smooth continuity.

overlay: (1) The superimposition of one image on another without the *background (4)* showing through. (2) A foreground piece in *animation*. It is registered on an upper *level* so permitting objects registered on a lower *level* to pass behind it without resorting to *match lines*. (3) Processing a computer *program* in instalments when it is too long for available memory in the system. Each segment overlays or replaces part of the previous code.

overload: (1) To apply too large a signal to a system. (2) To draw too much power from a supply.

overmodulate: To feed a signal into a system which has a greater *amplitude* than the normal input range of the system. **Figure O.6** shows the result for an optical sound track where the overmodulated peak signals are limited.

over-range: A situation where the input to a measuring instrument exceeds its display capacity.

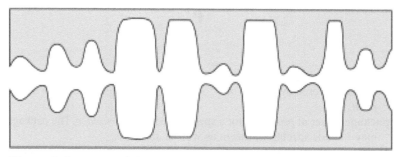

Figure O.6 Overmodulating an optical sound track

overscan: The increase in displayed television picture size needed to move the edges of the picture beyond the physical edges of the display device.

overshoot: A *transient* excess response at the beginning or end of a pulse signal. In a video display it appears as a light or dark line at the edge of an image where a marked change of *brightness* occurs (compare *ringing*). See **Figure O.7**.

Figure O.7 Overshoot

over the shoulder shot (OSS): A *reverse shot* of a subject taken over the shoulder of a person in the foreground.

oxide tape: Magnetic recording tape using ferric oxide particles coated in a *binder* on a flexible plastic *backing*.

P

Pacific Ocean region (POR): The area of the Earth centred on the Pacific Ocean covered by a range of *satellites (2)* in *geostationary Earth orbit*.

package: A set of *programs* for a specific task on a *computer*. The package may include special *hardware* as well as *software*.

packet: A block of data beginning with a *header* and with an optional *trailer (3)* at its end. It can have a fixed or variable length and may carry *error correction codes*.

packet identifier (PID): A code allocated to a video, audio or data *element (6)* of a programme in an *MPEG-2 transport stream*. Some PID values are reserved for carrying details of the *program specific information (PSI)*.

packetized: Describes a data stream which has been split into *packets*.

packet switching: Dividing a message into *packets* and sending them across a data *network*. *Packets* from different users may be interleaved. Also, the *packets* forming a message may not use the same route or arrive in the correct order .

pad: (1) A fixed value *attenuator* for video or audio signals. **(2)** In *computer graphics*, a *data tablet*.

pad roller: A small roller holding photographic film against a sprocket.

painting: Freehand drawing on a *computer graphics* display, where line width, opacity and other characteristics can be altered.

PAL: An analogue colour television system using *phase alternation by line*, with 625 lines per *frame (2)*, 50 *fields* per second and a 4.443 361 875 MHz *subcarrier* (see **Appendix R** for more details). It was proposed by Dr Walter Bruch in Germany in 1963 as an improvement of NTSC. It is used mainly in Europe, China, India, Australia and parts of Africa. See **Appendix I** for a full list.

PAL-B: A specific *transmission (1)* version of *PAL* to fit a 7MHz *channel (2)*. It has a *luminance (2) bandwidth* of 5 MHz and an analogue sound *carrier* spacing of 5.5 MHz. See **Appendix I** for countries.

PAL-G: A specific *transmission (1)* version of *PAL* to fit an 8MHz *channel (2)*. It has a *luminance (2) bandwidth* of 5 MHz and an analogue sound *carrier* spacing of 5.5 MHz. See **Appendix I** for countries.

PAL-I: A specific *transmission (1)* version of *PAL* to fit an 8MHz *channel (2)*. It has a *luminance (2) bandwidth* of 5.5 MHz and an analogue sound *carrier* spacing of 6 MHz. See **Appendix I** for countries.

PAL I–1: A variant of *PAL-I* with a smaller *vestigial sideband* of 0.75 MHz instead of the normal 1.25 MHz, used to minimize adjacent channel interference in parts of the UK *digital terrestrial television* transmission system.

PAL-M: A Brazilian analogue colour television system using *phase alternation by line*, but with 525 lines per *frame (2)*, 60 *fields* per second and a 3.575 611 49 MHz *subcarrier.*

PAL plus: A *widescreen* version of the *PAL* system giving compatible 16:9 *aspect ratio (1)* pictures to suitable equipped *widescreen* television *receivers.* See also *wide screen signalling.* (Hist.)

palette: In *computer graphics* and *vision mixers*, the range of pre-determined colours from which a selection may be made.

pan: (1) To rotate a camera in a horizontal direction. (2) In *animation*, movement of the field in any direction. (3) Positioning the apparent source of sound within the area of a stereo sound presentation.

pan-and-tilt head: A *mount* for a film or video camera on its *tripod* or *pedestal* which allows smooth rotation in both horizontal and vertical planes.

Panaglide: A harness with a stabilized *camera mount* giving smooth movement of a hand-held camera. (T rade name.)

Panavision: The developers of a set of widescreen film standards. In Panavision, 35mm film is used with an *anamorphic (1)* lens on both the film camera and projector. In *Super-Panavision*, 65mm film is used in the camera, but prints are made on 35mm film by *anamorphic* reduction. In *Ultra-Panavision*, 65mm film is used in a camera with an *anamorphic* lens to give a 1.25 :1 reduction.

pancake: (1) A reel of tape on a *core (1)*. (2) A low platform to raise artistes or objects off the studio floor to improve composition.

panchromatic: Having a response to all the colours in the *visible spectrum.*

Pandata: A system allowing camera lens information such as zoom, focus and *T-number* to be digitally encoded and saved, for example on a camcorder's digital audio channels. (*Panavision* trade name.)

pan handle: See *panning bar.* (Colloq.)

panning bar: A bar fitted to a *pan-and-tilt head* to enable smoother and more precise control. Also called a *pan handle*, particularly in the USA.

panorama: In *slide* projection, a single complete picture covering the whole or several parts of a *multi-screen* presentation.

pan pot: A *potentiometer* control for positioning the apparent source of sound within the area of a stereo sound presentation.

pantograph: A lighting suspension system widely used in television studios with a scissors type construction allowing *luminaires* to be easily raised or lowered. Also colloquially called a *lazy boy.*

PAR: A lamp with an in-built *parabolic reflector.*

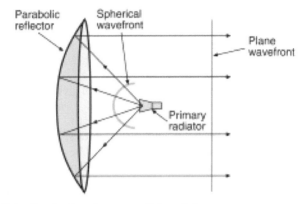

Figure P.1   Parabolic microwave dish aerial

parabolic reflector: A concave surface whose cross-section along the axis is a parabola and which can be used to focus light, sound or *radio frequency* radiation. See **Figure P.1**.

parade display: Display on a *waveform monitor* of *component video* signals side-by-side. See **Appendix Z** for an example.

parallax: The apparent shift in object positions resulting from a change in viewpoint. In film camera work, the difference between the composition of the objects in the picture as seen through the *viewfinder (2)* and through the camera lens.

parallel interface: A multi-wire electrical connection between parts of a system, e.g. the *parallel port* on a *personal computer* or a *SCSI bus*.

parallel port: A bidirectional high capacity *parallel interface* found on IBM compatible *personal computers*. It is also known as the enhanced *Centronics interface*, and is defined in the *IEEE* 1284 standard.

parameter: A number or value used in the specification of an electronic system.

parametric equalizer: An audio *equalizer* in which the gain, frequency and *bandwidth* of the filter can be varied.

paraxial focus: The *focal point* of a lens for rays close to the *optical axis* in the absence of *spherical aberration*. See **Figure S.7**.

parcan: A simple, low cost *luminaire* in the form of a metal tube for use with a *PAR* lamp.

parity: In digital systems, an additional check *bit* added to a data *word* to enable a single bit error to be detected in each *word* when the data *words* have been transmitted or recorded. In odd parity , the state of the parity bit is set to make the total number of ones in the *word* an odd number. In even parity, the total number of ones in the *word* is made even. A parity checker in the receiving device adds up the number of ones in the incoming data *word* and indicates a data error if the result is not a proper odd or even total.

partitioning: Dividing a large capacity physical *hard disk drive* into several smaller virtual disks.

parts per million (ppm): A measure of the concentration of one substance mixed with another.

pascal (Pa): *SI unit* of pressure, equal to one *newton* per square metre. Named after the seventeenth century French scholar and scientist Blaise Pascal.

passive component: An electronic component which does not contain a source of power.

paste up: An assembly of different visual or text components on to a common base to form a coherent whole.

patch: **(1)** A transparent piece of thin film used to repair a break in a photographic film. **(2)** A small *program* inserted to remedy a defect in a larger *computer program*. **(3)** To temporarily make a connection on a *patch panel*.

patch lead: A short cable with plug at each end used with a *patch panel* to provide a (usually) temporary interconnection.

patch panel: An arrangement of one or more *jackfields* which allows different outputs to be connected to different inputs using *patch leads* or *U-links*.

pause: **(1)** A temporary interruption to a programme. **(2)** Stopping a *magnetic tape transport* mechanism without changing the operating mode (record or play back). In a broadcast quality *videotape recorder* with *dynamic tracking* it will give a broadcastable still picture (compared with *stop* which will not).

pause control: A control on a tape recorder which temporarily stops the tape *transport* without unthreading. It may or may not give a steady still picture on a *videotape recorder.*

pay per view (PPV): A television distribution system where a one-off payment is charged for viewing a specific programme or event.

pay TV: A television distribution system where a regular subscription is charged for viewing the programme services provided.

payload: **(1)** In a *satellite (2)* launch rocket, the *satellite (2)* is the payload. **(2)** In a data *network* system, the user's data is the payload.

Pb: *Colour difference signal* used in *analogue component* video. Pb = 0.56433 (B–Y). See **Appendix Z**.

PC-card: An *interface (1)* for *personal computer* expansion modules, often used with laptop portable computers because of its slim profile. It is also the standard adopted by *DVB* for detachable *conditional access* modules using the *DVB* Common Interface.

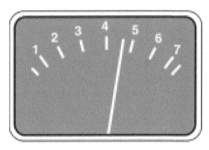

Figure P.2   PPM scale

peak amplitude: The *amplitude* of a signal at its highest level.

peak programme meter (PPM): An audio signal level meter which monitors the peaks of the applied signal. See **Figure P.2**.

peak-to-peak (p-p): The *amplitude* of a signal measured from its most negative level to its most positive level.

peak white: The maximum allowed *amplitude* of a video signal corresponding to the brightest white area in the scene.

pearlescent: Cinema *screen (1)* material with a slightly translucent coating.

ped: See *pedestal (1)*. (Colloq.)

pedestal: (1) A wheeled steerable *camera mount* with adjustable height which uses a compressed gas piston to make it easy to raise or lower the camera. **(2)** In a North American 525-line analogue video signal, a 7.5 *IRE* units fixed rise in *black level* above the *blanking level*. There is no pedestal in 625-line television systems or the 525-line system used in Japan. Called *set up (5)* in the USA. See **Appendix K**.

ped up, ped down: Movement of a video camera vertically up or down on a *pedestal (1) camera mount*.

peer-to-peer network: A *network* with no dedicated *servers*, sometimes called a *workgroup*.

peg animation: *Animation* in which the illusion of motion is provided by substituting artwork every *frame (1)* or every second *frame (2)*. The artwork is registered by pegs.

peg bar: Metal pins allowing artwork and *captions* to be accurately registered.

pencil test: See *line test*.

Penguin: A track mounted camera *dolly* made by Vinten and sometimes used with a *Dolphin arm*. (Trade name.)

perceptual coding: Using shortcomings of the human visual system or hearing to reduce the amount of data needed to convey a digital video or audio signal.

Perceval: Belgian teletext system.

perforated screen: A projection *screen (1)* with small holes to permit a *loudspeaker* system to operate efficiently from the rear side.

perforations: Holes along the edge of a strip of film used for its *transport* and *registration (1)*. Also colloquially known as *sprocket holes*.

Performing Rights Society Ltd (PRS): A UK association of composers, authors and publishers of musical works which collects public performance and *broadcast* royalties on behalf of its members. (Website: www.prs.co.uk).

perigee: The point in a *satellite's (2)* orbit when it is nearest to the Earth.

period: The length of time it takes a wave to complete one cycle. The period equals 1/*frequency*.

peripheral: An external device added to a *computer* system, e.g. *floppy disk* drive, printer, *visual display unit*, etc.

Peritel: See *SCART connector*.

Perlux: A projection *screen (1)* material of high reflectivity over a wide angle. (Trade name.)

persistence: The time taken for an image (usually on a *cathode ray tube*) to die away.

persistence of vision: In human vision, the short time for which an image remains on the *retina*. It allows us to see movement in films and television as continuous motion.

personal computer (PC): (1) General term for a small desk-top *computer* system using a *microprocessor*. **(2)** The desktop computer launched in 1981 by IBM. It had a 4.77 MHz Intel 8088 *microprocessor*, 16 kilobytes of *RAM* and one 13.3 cm (5.25 inch) *floppy disk* drive. It sold for $1565 (about £1000).

Personal Computer Memory Card Interface Association (PCMCIA): The group which defined the *PCMCIA interface (1)* which is now called *PC-card*.

personal video recorder (PVR): A disk-based digital video recorder inside a *set top box*.

perspective control lens: An *objective lens* designed to be moved off-axis so that more than one *projector* can produce an image free from *keystone distortion* on the same part of a *screen (1)*. Also called an *offset lens*. See **Figure P.3**.

petabyte: 1024 *terabytes*.

petal: A segment of a dismantled *SNG dish (2)* aerial.

P-frame: Predictively coded frame. An *MPEG* compressed picture in a video sequence which can take into account redundancy between itself and the previous *I-frame* or *P-frame* in the *GOP* sequence. *Macroblocks* of the picture to be coded are compared with the corresponding ones

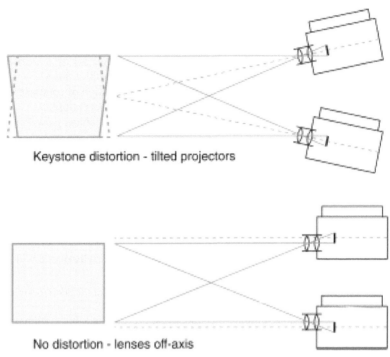

Keystone distortion - tilted projectors

No distortion - lenses off-axis

Figure P.3   Perspective control lens

(which may be shifted using *motion estimation*) in the previous picture to find the one with least changes (i.e. most *redundancy*). This makes the P-frame less efficient at coding images than a *B-frame* since it only looks back at the previous *frame (2)* and not forwards to the next in its search for *redundancy*.

phantoming: A technique for sending more than one signal along a pair of wires in a cable. See *phantom power*.

phantom power: Power fed to a *microphone* using the audio signal cable without interfering with the audio signal.   Also used in video camera cables.

phase: (1) The timing relation between two signals of the same frequency. If the time for a cycle of one signal is represented as 360 degrees along a time axis, the relative timing difference of a corresponding point on the second signal is called the *phase difference* or *phase error* and is expressed in degrees.   When the signal timings coincide exactly the signals are '*in phase*'. **(2)** The angle between two vectors. See also *phase modulation* and **Figure P.4**.

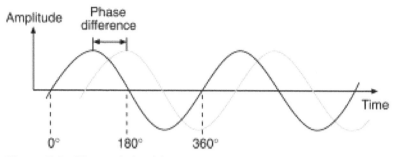

Figure P.4 Phase relationship

phase alternation by line (P AL): Inverting the *phase* of the *V colour difference signal* from one line to the next to minimize the perceived hue shifts resulting from *phase* errors introduced in transmission. Hence name of the colour television coding system. See also *PAL* and *PAL-M*.

phase control: (1) A control to rotate the display on a *vectorscope*. **(2)** A method of altering the moment of switching a *thyristor* or *triac*, relative to the phase of the supply mains, to provide power control, e.g. for lamp dimming in a studio.

phase difference: See *phase (1)*.

phase error: See *phase (1)*.

phase jitter: Spurious changes in the *phase (1)* of a signal.

phase-locked loop (PLL): An electronic circuit in which an oscillator is controlled using feedback to either lock its output to track changes in a varying input signal or remove *phase jitter* from a signal by locking to the average *phase (1)* of the signal, like a flywheel. In digital television systems they are used extensively to generate *clock* signals a reference video such as *colour black*.

phase modulation: *Modulation* where the phase angle of a *carrier signal* is varied, e.g. to transmit the colour component in the *PAL* video signal. In a digital system, a number of fixed phase angles are used, called *states*. See also *phase (1)*.

phase shift: A movement in the relative timing between two signals. It is measured in degrees, where a 360 degree phase shift corresponds to a complete cycle of the reference periodic waveform. See also *phase (1)*.

phasing: (1) Audio special effect created by recombining a split signal after delaying one part by a very small amount. Varying the delay alters the frequency-dependent phase cancellations to create the ef fect. **(2)** Checking that *loudspeakers* are *in phase* with one another (rather than in *anti-phase*). **(3)** Adjusting the timing of an analogue video signal to match a reference signal.

Phenakistoscope: A viewer using a pair of parallel discs revolving on a shaft. Drawings on one disc were viewed through slits in the edge of the second to give the illusion of movement. It was invented in 1833 by a Belgian, Joseph Antoine Plateau. (Hist.)

phon: Unit of perceived loudness, on a scale where 0 phon level is taken as the threshold of hearing and is defined to correspond to a *sound pressure level* at 1 kHz of 20 µPa (micropascal). At other frequencies the loudness of a sound in phons is equal to the intensity in *decibels* of a 1 kHz tone judged to be equally loud.

phono plug: A small *coaxial cable* single-pin audio or video connector , also called an *RCA connector.* (Trade name.)

phosphor: A specific substance which can be made to emit visible light of specific *wavelengths* when irradiated, as for example by an electron beam in a *cathode ray tube* or by *ultra-violet* radiation in a fluorescent lamp.

phosphor decay: The time taken for an image on a *cathode ray tube* to die away after the *electron gun* is switched off.

Photo CD: A *CD* format containing up to 100 digitized photographic images at a variety of resolutions. It is defined in the *Orange Book.*

photocell: An electronic device which converts light into an electrical signal.

photodiode: A semiconductor device used to convert light into variations in an electrical signal.

photo-electric cell (PEC): A device for converting light into an electrical signal.

photographic sound: System of recording and reproducing on motion picture film in which the sound track takes the form of variations in the *density* or width of a photographic image.

photo-mechanical transfer (PMT): A term used to describe the reproduction of existing material. (Hist.)

photometer: An instrument used to measure the luminous intensity of light sources.

photomultiplier: A very sensitive *vacuum tube* photoelectric detector with built-in high gain amplification and used in some *telecine* machines.

photopic curve: The overall response of the human eye to electromagnetic radiation, established by the *Commission Internationale de l 'Éclairage* in 1924. It is properly known as the photopic spectral luminous efficiency function (compare *scotopic curve*). See **Figure P.5**.

photosensitive: Sensitive to light.

phototransistor: A transistor able to convert light into an electrical signal with amplification.

pick: The selection of a display item in *computer graphics*.

pick-up (PU): A *transducer* for reproducing recordings on vinyl *audio discs*.

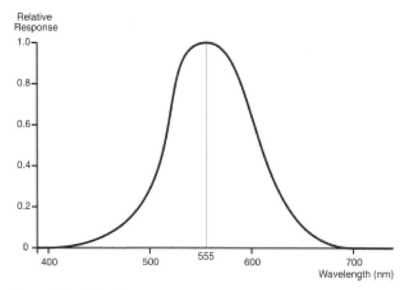

Figure P.5 Photopic curve

pick-up arm: A device for supporting a *pick-up*.

pick-up tube: A video camera *tube (2)*. (Hist.)

picosecond (ps): One thousandth of a *nanosecond*.

picture element: See *pixel*.

picture in picture (PIP): The insertion of a reduced size image of another television channel into the main viewed picture on a suitably equipped television receiver.

picture line-up generating equipment (PLUGE): A video test signal used to set contrast and *brightness controls* on picture monitors. To set black level, the *brightness control* on the monitor should be set so that the super black bar is not visible but the bar slightly lighter than black is just visible. See **Figure P.6**.

picture search: The ability to locate a *frame (1)* on film, *videotape* or *disk* over a range of speeds.

picture tube: See *cathode ray tube*.

pictures per second (pps): An alternative description of a film or video *frame rate*.

piece to camera (PTC): A reporter giving a news report face-on to the camera.

piezo-electric: Producing an electrical voltage by mechanical deformation of a material, usually a crystal (e.g. a crystal microphone), or using a

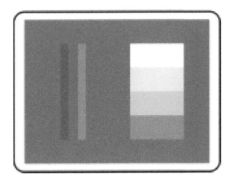

Figure P.6 PLUGE picture display

voltage to deform a material, e.g. a crystal earpiece or *dynamic tracking videotape* replay *heads*.

pillarbox: A television picture presentation on a *widescreen* display with black bars at the left and right sides giving a narrower viewed *aspect ratio (1)* (compare *letterbox*).

pilot pin: A pin in a motion picture or *rostrum camera* or printer mechanism, engaging a *perforation* hole for precise *registration* during *exposure*. See *register pins*.

pilot tone: A speed control/camera *sync* system. The tone may be recorded on a full width tape using two narrow tracks in *anti-phase*.

pin: The component of a film camera or *projector* mechanism which engages in the *perforation* hole to locate the *frame (1)*.

pinch roller: See *pressure roller*.

pincushion distortion: Image distortion in an optical or video system causing a rectangle to appear to have concave sides and extended corners (compare *barrel distortion*). See **Figure P.7**.

pinhole: A small defective spot in a photographic emulsion or image, or in a magnetic coating.

pinhole slide: A *slide mount* with a small hole in the middle, used in *projector* lamp and mirror adjustment by forming an image of the filament and its reflection on a translucent lens cap.

pink noise: *White noise* filtered at 3 dB per octave to give constant energy per octave bandwidth.

pin registration: The exact location of a film *frame (1)* or image, as in a camera gate or *slide mount*, by means of pins engaging the *perforation* holes.

pipe: A data circuit. (Colloq.)

pipping: The copying of data from one *computer* disk to another.

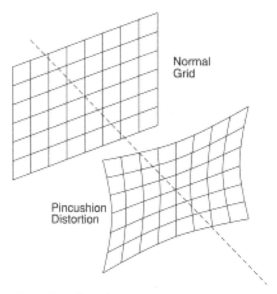

Normal Grid

Pincushion Distortion

Figure P.7   Pincushion distortion

pitch: (1) The distance between successive regularly occurring points, such as film *perforations* or video recorded tracks. **(2)** The frequency of a sound within the audible range.

pitch changer: A device to alter the *pitch* of an audio signal. It is used with a small change of *pitch* to reduce the possibility of *threshold howl*.

pitch control: A narrow range variable speed control on a magnetic or *audio disc* reproducer which can allow the replay to be exactly tuned to a musical instrument.

pixel: Abbreviation for *picture element*. The smallest unit which makes up a video picture. The name is often applied to one *sample* of digital picture information.

pixellation: (1) A visual effect in film and video where moving action is represented as a series of stills each held long enough to be recognized as static. **(2)** Also in video, an effect in which the whole picture is reproduced as a comparatively small number of enlarged *pixels*. See *tile*.

plasma panel: A display system, comprising a matrix of gas-filled cells which can be turned on and off individually. It forms the basis for large flat screen video displays up to about 1.5 metres across.

plate: A photographic print, motion picture film or still *transparency*, used as the *background (3)* in special effects such as *back projection*, *reflex projection*, *aerial image photography* and *travelling matte* shots.

platen: (1) The glass pressure plate for *cels* and *background (4)* on an *animation stand*. (2) The working surface of an *overhead projector*. (3) The paper supporting surface, often cylindrical, on a *computer* printer.

platform: In computing, the *hardware* system designed around a particular type of *central processor unit*.

platter: (1) A large horizontal disc used to support long rolls of film in *continuous projection* systems. (2) A non-magnetic rigid disc coated with a magnetic material in a *hard disk drive*, often made from aluminium or glass/ceramic. See **Figure H.1** and **Figure H.2**.

playback: The reproduction of a recording.

plotter: A device using coloured pens to draw graphical hard copy from a *computer*. See also *X-Y plotter*.

Plumbicon: A photoconductive lead oxide vidicon camera *tube (2)* made by Philips and launched in 1962. (T rade name.) (Hist.)

podium: A dais or raised platform used by a presenter or lecturer .

point: (1) The *exposure* increment used in a motion picture printing machine. A printer point scale of 1 to 50 is normally used. (2) In typography, the height of characters expressed in units of 0.353 mm (exactly $\frac{1}{72}$ of an inch).

point of pr esence (PoP): The physical location where a long distance *telecommunications* company's lines terminate and connect to a local *telecommunications* company or directly to customers.

point-to-point: A connection between two individual pieces of equipment or *nodes (2)* on a *network (1)*.

PO jack: Post Office jack. A three circuit *jack* plug of 6.25mm ($\frac{1}{4}$ inch) nominal diameter. See **Figure P.8**.

Polacoat: A material for fine-grain rear-projection *screens (1)*, available in glass, rigid plastic, or flexible plastic. (T rade name.)

pola filters: Small sheets of *Polaroid*, sometimes mounted in glass, which can be fitted in front of a camera lens.

polar diagram: A plan view of the response pattern of a *microphone*, *loudspeaker*, or *aerial* system. Also applicable to light sources and projection *screen (1)* reflection or transmission characteristics. See **Figure P.9**.

polarization: Modification of an *electromagnetic wave*, especially light, so that its oscillation takes place in a plane or a limited series of planes. See **Figure E.1**.

polarity: Of a positive or negative electrical state.

polarizer: In consumer *satellite (2)* signal reception, an electromechanical or electromagnetic device fitted between the source and the *low-noise block* at the focus of the *aerial* to rotate the plane of *polarization* to match that of the *low-noise block's* probe.

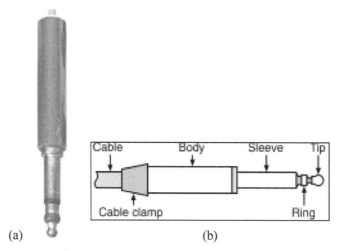

(a)   (b)

Figure P.8   PO jack

Polaroid: (1) A transparent plastic material capable of *polarizing* visible light. (Trade name.) **(2)** An instant-picture photographic system. (Trade name.)

polar orbit: A *satellite (2) orbit* which passes over both poles of the Earth.

Polascreen: A transparent, neutral coloured *filter (1)* which *polarizes* transmitted light. (Trade name.)

polecat: A spring-loaded pole which can be mounted between floor and ceiling for rigging on location.

pole tips: The parts of a magnetic video *head (2)* which protrude beyond the circumference of the *head drum* to make contact with the tape.

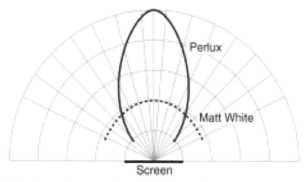

Figure P.9   Polar diagram of screen reflectivities

popping: An effect undergone by some card mounted *slides* subjected to projection lamp heating causes distortion of the film and an image on the *screen (1)* which pops in and out of focus.

port: A hardware connection point to interconnect a computer and other hardware *peripheral* devices or networks, e.g. an *Ethernet* port or a video port.

portable single camera (PSC): A technique of television programme production having much in common with film shooting methods and employing a single television camera and video recording equipment, often combined in a *camcorder.*

portapack: A portable battery-operated video camera and recorder . (Hist.)

portholing: (1) A circular shading ef fect caused by limitations in the optical performance of the lens on a film or video camera. See *vignetting.* **(2)** A circular shading ef fect caused by incorrect beam alignment in video cameras fitted with pick up *tubes (2)* rather than *CCD imagers.* (Hist.)

portrait: A picture format where the image has greater height than width (compare *landscape*).

positive: A photographic image in which the tones (and colours where they appear) reproduce those of the original scene (compare *negative*).

positive feedback: A condition where some of the output of a system is connected back to the input *in phase* with the input. This is used to advantage in some oscillator circuits, but when it occurs unintentionally there can be a build-up in *amplitude* leading to instability or unwanted oscillations in the system, e.g. *loudspeaker* output coupled to a *microphone* input produces undesirable *howl-round.*

positive temperature coefficient (PTC): A substance or component whose electrical resistance rises with rising temperature.

posterization: A digital video effect where the image is reproduced using only a few specific tones or flat colours, with most of the image gradation and *detail* suppressed.

post production: The *editing* of pre-recorded material, including use of special effects and audio *dubbing.*

post sync: Post-*synchronization.* Recording synchronous dialogue or sound effects for a scene after it has been photographed.

post, telegraph and telephone (PTT): Usually refers to the *tele-communications* authority within a country. See also *common carrier.*

potential difference (PD): The voltage difference between two points in an electrical circuit.

potentiometer: A variable resistor where the *wiper* and the two ends are employed as a potential divider .

pounds per square inch (PSI): An imperial measure of pressure, still used for car tyre pressures in the UK.

POV shot: Point-of-view shot, as though seen by the actor .

Power Amplifier (PA): An electronic *amplifier* which provides a high power output, commonly used in audio for driving *loudspeakers*.

power factor: A measure of the *phase* relationship between voltage and current in an AC transmission circuit. If the current and voltage are out of phase the power factor will be low and there will be transmission losses. Unity power factor represents a non-reactive *load (2)*.

Pr: *Colour difference signal* used in *analogue component* video. Pr = 0.71327(R–Y). See **Appendix Z**.

Praxinoscope: An early *animation* device using a mirrored drum, designed in 1877 by Emile Reynaud, a French inventor .

preamble/postamble: In digital recording, *sync* and identification groups recorded before and after each data *block (1)*.

preamplifier: A preliminary stage of amplification to boost a low-level signal before further amplification is applied.

pre-echo: Unwanted hearing of an audio signal before it should have arrived. It can arise from adjacent grooves of *audio disc* records, or from magnetic transfer between adjacent turns of wound *magnetic tape*. See also *print through*.

pre-emphasis: A means of boosting the gain of an audio signal over a particular frequency range before recording or transmission to improve the overall *signal-to-noise ratio*.

pre-fade listen (PFL): 'Piffle'. A facility on a sound desk, which allows channels to be heard before *fading up*.

premiere: The first major public presentation of a motion picture, video or multiscreen production.

premix: A mix of several sound components prepared before the final mixing operation.

pre-recorded: Material for inclusion in a programme that is already in a recorded form.

preroll: In editing, a specific amount of time allowed for a *videotape recorder* to run before an edit to get it up to speed and *synchronized* for the edit.

presence: Boosting the *frequency response* of an audio circuit, between 3 kHz and 8 kHz, to create an illusion of nearness to the sound source.

presentation time stamp (PTS): In *MPEG-2*, an indicator of the display time for a decoded *MPEG-2* picture *frame (2)* or audio *frame (5)* (compare *decoding time stamp*).

presentation unit: *MPEG-2* term for an uncompressed video *frame (2)* (compare *access unit*).

Presfax: BBC communication service using *ICE* to provide regions with programme scheduling information. (Hist.)

Press Association (PA): UK national new agency supplying news, sport and information services to all national and regional newspapers, major broadcasters, online publishers and other commercial or ganizations. (Website: www1.pa.press.net).

pressure pad: A softly sprung felt component which ensures good contact between the surface of *magnetic tape* and the face of a magnetic *head (2)*.

pressure plate: The part of a motion picture camera, printer or *projector* which holds the film flat at the time of *exposure* or projection.

pressure roller: A wheel, usually made of rubber , which holds the tape against the *capstan* in a *magnetic tape* recorder mechanism.

Prestel: *British Telecom's* UK *viewdata* service. (Trade name.)

Presto: A multi-channel digital audio interface developed by the *Audio Engineering Society (AES)* , based on *Synchronous Digital Hierar chy (SDH)* at data rates between 155 Mbit/s and 10 Gbit/s.

preview: (1) In video, to see a picture before transmission or recording. **(2)** A special presentation of a completed motion picture production prior to its public exhibition. **(3)** In video editing, to rehearse an edit without recording it. **(4)** An output of a *vision mixer (1)* allowing a picture or action to be seen or rehearsed on a picture monitor .

preview monitor: A picture monitor in a television studio centre control room to allow a picture source to be viewed before transmission or recording.

primary: The input winding of a *transformer.*

primaries: See *primary colours.* Also used to denote the phosphors on a colour *cathode ray tube* picture display.

primary battery: An assembly of *primary cells.*

primary cell: An electrochemical *cell* which generates electrical energy by chemical changes which eat away one of its electrodes. The process is not reversible and the cells generally cannot be rechar ged.

primary colours: A number of colours (usually three) in a colour reproduction system from which all available *hues* can be produced by mixing. In an *additive colour* process, such as colour television, they are red, green and blue. In a *subtractive colour* process, such as colour film, they are yellow, *magenta* and *cyan.*

primary radiator: Microwave horn *aerial* at the main focus of a *parabolic reflector.* See **Figure P.1.**

prime lens: An *objective lens* of a camera or *projector* of normally fixed *focal length.* Supplementary attachments or a range extender can be added.

principal plane: The plane through a *principal point* of a lens at right angles to its *optical axis.* See **Figure P.10.**

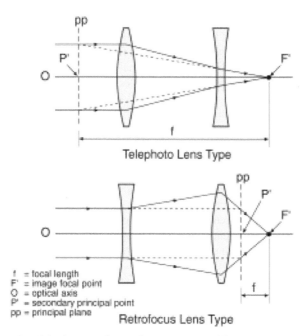

Figure P.10   Telephoto and retrofocus compound lenses

principal point: The point on the *optical axis* of a *compound lens* which would be the centre of the equivalent simple lens with the same *focal length*. There are two principal points, one for the object *focal point* (primary) and one for the image *focal point* (secondary). See **Figure P.10**.

print: A photographic copy of a film, usually with a positive image.

printed circuit board (PCB): An insulating board on which the interconnections (tracks) for an electronic *circuit* have been printed or etched.

printer: (1) A *computer peripheral* device to produce *hard copy*, i.e. on paper. **(2)** Machine for the *exposure* of film to produce photographic copies.

printer lights: Figures representing the *exposure* levels required in a motion picture printer, usually quoted for red, green and blue on a scale of 1 to 50.

print server: A *server* dedicated to controlling the flow of print jobs to a printer.

printout: The printed output of a *computer*. Also called *hard copy*.

print through: Unwanted transfer of the signal on a *magnetic tape* record to adjacent layers of the tape when wound on a *spool*.

prism: A transparent solid with inclined flat surfaces which re-directs a light beam by *refraction* and/or internal reflection.

private tables: Supplementary data tables about the digitally transmitted programme services which are defined and used in *DVB* in addition to the *MPEG–2 PSI* tables.

probe: A tool used to gain access to signals in an electronic unit under test, with minimum disturbance.

proc amp: A processing amplifier, used to stabilize an analogue *composite video* signal. It regenerates and replaces the *sync pulses (2)* and *colour burst*, re-blanks the signal and may allow adjustment of *luminance (2)* gain, *chroma (1)* gain, *black level*, etc.

process shot: (1) A general term in motion picture production for a trick shot created by special effects photography and/or optical printing. **(2)** A studio shot in which the *background (3)* is a still or moving projected picture.

processing: The chemical treatment of exposed photographic material so as to render the *latent image* permanently visible.

proctor: A large wheeled *camera mount* with rubber tyres for use on rough ground.

Producers Alliance for Cinema & Television (PACT): Founded in 1991, it is the only UK trade association representing independent television, feature film, *animation* and new media production companies. (Website: www.pact.co.uk).

production assistant (PA): A general assistant to the director or producer, responsible for scripts, *continuity* and logging of shots.

production master: A modified copy of a film for production use.

program: (1) A sequence of instructions which executes in a *computer* when it is run. Also called *software*. **(2)** In *MPEG*, a program is a broadcast television or radio channel, e.g. BBC 1 or Capital Radio in the UK. See also *service (2)*.

program clock r eference (PCR): A count carried within an *MPEG-2* digital *multiplex (2)* to enable the decoder reference clock to lock to the encoder/*multiplexer (2)* reference.

programme: The show material to be seen and heard when the production is presented.

program allocation table (P AT): A data table carried in an *MPEG-2* digital *multiplex (2)* which identifies the *PMTs* for each of the programmes in the *multiplex (2)*. It is part of the *MPEG-2 programme specific information*.

programmable array logic (P AL): An integrated circuit whose logic element interconnections are programmable for a specific application.

program map table (PMT): A data structure carried in an *MPEG-2* digital *multiplex (2)* which contains a list of *PIDs* and other descriptors which identify the video, audio and data services that make up a single programme carried in the *multiplex (2)*. It is part of the *MPEG-2 program specific information (PSI)*.

programmed instruction: A method of instruction in which the student learns by following a controlled sequence of instructions.

programming: Creating *software* to instruct a *computer* what task to perform and in what order to do it.

program specific information (PSI): Data carried in an *MPEG-2* digital *multiplex* which identifies each of the *services* being carried in the *multiplex (2)*. It is in the form of four types of data table: *PAT, PMT, CAT* and *private tables*.

program stream (PS): An *MPEG-2* digital *multiplex (2)* of video, audio and data associated with a single *service* and optimized for storage and retrieval of material in a *quasi error free* environment, e.g. *Digital Versatile Disc (DVD)* . It uses very long packet lengths (compare *transport stream*).

progressive scanning: A video *scanning* system in which all lines comprising a *frame (2)* are scanned or drawn in sequence rather than *interlaced*. Also called *sequential scanning* (compare *interlace*).

projection television: A video presentation system in which the picture is optically displayed on a separate large *screen (1)*, rather than on the face of a *cathode ray tube*.

projector: A device to project images on to a *screen (1)*, from a motion picture film, transparent photographic *slides* or strips, or electronically generated video sources.

projectors: In *computer graphics*, lines passing through an object to intersect with the view plane of projection.

projector stack: A stand holding several automatic *slide projectors*, usually one above the other .

prompt: A character or string of characters on a display which lets the user know what action is expected.

prompter: A device used to provide a visible script for on-camera presenters.

prop: Stage or studio property . Any item used in a production except scenery and costumes.

proportional spacing: In *computer graphics* or printing, adjusting the separation of characters forming the text on a *computer screen*, or being printed, to give a balanced spacing where each character does not occupy the same width.

props: Scenery property such as furniture, food, guns, newspapers, etc.

protocol: (1) A procedure agreed upon by agencies wishing to communicate. (2) In data *networks*, a strict set of rules for physical connection, transmission mode, data rate and format, and error handling on the *network*. See also *Open System Interconnect (OSI)*.

protocol stack: In networks, the various *protocols (2)* which allow successful data communication to take place. See also *Open System Interconnect (OSI)*.

psophometer: A *noise* measuring instrument.

psychoacoustic: In audio coding, using an understanding of human audio perception to encode an audio signal. These techniques are used in some audio data compression systems, e.g. *MUSICAM, MPEG* audio and *ATRAC*.

public address: An audio system for presenting speech or music to a large audience.

puck: In *computer graphics*, a hand-held input device used on a *data tablet* to accurately enter co-ordinates by the use of programmable buttons and sometimes a cross-hair guide.

pull back: The backward movement of a film or video camera away from its subject.

pull-down: The operation of moving film from one *frame (1)* to the next in a camera or *projector* mechanism.

pull list: A list of negatives to be pulled from store in negative cutting.

pulse: A rapid signal change from one voltage level to another and back again. It is usually of short duration and can include a short burst of tone. In this case, as an *audio-visual* cueing signal it is properly known as a *cue tone*.

pulse and bar: A video test signal with the displayed appearance of a thin vertical white line (the *pulse*) to the left of a large white block (the bar). It is used to identify and measure *linear distortions* of the television signal. See **Appendix L**.

pulse-code modulation (PCM): A *time division multiplex* technique where an analogue signal is *sampled* and *quantized* at regular intervals into a digital signal, then represented as a coded arrangement of *bits*.

pulse cross: A means of displaying the normally invisible *sync* signals and *blanking* areas on television picture display equipment.

pulse distribution amplifier (PDA): A video timing pulse *amplifier* designed to provide several separate output signals. It may also perform cable equalization.

pulse position modulation (PPM): A signalling system in which information is carried by varying the position of pulses within a constant repetition time period.

pulse to bar ratio: One of the video signal tests carried out using a *pulse and bar* signal. It is the ratio of the pulse amplitude to the bar amplitude at its centre expressed as a percentage and indicates the relative response of the system being tested to high and low video frequencies.

pulse-width modulation (PWM): A signalling system in which information is carried by varying the width of pulses of constant repetition rate.

pumping: See *breathing (1).*

pup: A small spotlight.

purity: A measure of the accuracy with which particular colours are portrayed on a picture monitor .

Purkinje effect: The perceived shift of maximum sensitivity of human vision towards blue in low light levels, first noted by a Bohemian physiologist, Johannes Evangelista von Purkinje, in 1825.   Also called the Purkinje shift. See also *scotopic curve.*

push-on, push-off: (1) A video *transition* in which the first scene appears to be pushed horizontally of f the *screen (1)* by the second scene (compare *conceal/reveal, scroll*). **(2)** A type of electrical switch.

push processing: See *forced development.*

Python: A counterbalanced 5 metre long camera *boom* arm which can be fitted to various wheeled or fixed supports. (T rade name.)

Q

Q-factor: A measure of the efficiency of a resonator or resonant circuit. High Q values represent a narrow width of *resonance*. Also called *magnification factor*.

Q-signal: In the *NTSC* colour system, represents a *colour difference signal* on the green-*magenta* axis (compare the *I-signal*).

Quad: Abbreviation for *Quadruplex*. (Hist.)

quad-in-line (QIL): A type of *integrated circuit* package with the contact pins in two pairs of offset staggered rows for mounting through holes in a circuit *board (1)*.

quadraphony: A four channel sound system which allows the reproduction of left-right and front-back sound distribution to a listener . (Hist.)

quadrature: A 90 degree *phase difference* between two signals.

quadrature amplitude modulation (QAM): Digital modulation of two signals with 90 degree *phase difference* derived from the same *carrier*. It results in a *constellation (1)* of phase/amplitude combinations. With 8 options for each of the two modulating signals, the resulting *constellation (1)* would have 64 possible values and is known as 64-QAM.

quadrature phase-shift keying (QPSK): *Phase modulation* with four states used in some data communications systems.

Quadruplex: An obsolete *open reel* analogue *videotape recorder* format using 50.8 mm (2 inch) wide tape with four record/replay *heads (2)* mounted on a rotating *headwheel* and using *transverse scanning*. (Trade name.) Also called '2 inch'. (Hist.)

quad split: A video effect where the picture is split into four boxes, each containing a different video source.

Quantel: QUANtized TELevision. A UK company specializing in the manufacture of digital *computer graphics*, *editing* and storage equipment. (Website: www.quantel.com).

Quantel digits: A proprietary *Quantel interface (1)* standard. (Hist.)

quantizing: The process in converting an *analogue signal* to digital form which takes place after *sampling*. The amplitude of each sample is rounded to the nearest of a defined number of discrete levels and assigned a digital value. See **Figure Q.1**.

quantizing error: Inaccuracy in the digital representation of an analogue signal due to the limited number of discrete steps into which the signal is analysed. See **Figure Q.1**.

quarter common intermediate format (QCIF): A video picture format with 176 *pixels* by 144 lines at up to 30 *frames per second*, used for video *telephone* applications. See also *common intermediate format*.

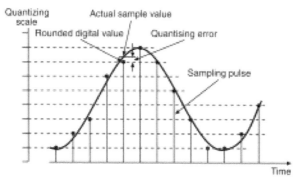

Figure Q.1   Sampling and quantizing in analogue to digital conversion

quartz crystal: A slice of quartz cut to resonate at a stable defined
frequency under the influence of an electric field using the *piezoelectric*
effect.

quartz halogen lamp: See *quartz iodine lamp*.

quartz iodine lamp: A small efficient lamp in which the tungsten filament
is enclosed in a quartz bulb containing an inert gas and a trace of a iodine
(a halogen). Also known as *quartz halogen lamp* and (less accurately)
quartz lamp. See also *tungsten-halogen lamp*.

quartz lamp: See *quartz iodine lamp*.

quasi error-free (QEF): Having a very low *bit error rate* (less than 10^{-10}).
The term can be applied to digital storage medium such as CD, DVD,
hard disk drive or some types of transmission channels.

qwerty: The standard English typewriter keyboard layout, named from the
first six characters of the top row of letters.

R

rack: (1) See *equipment rack.* (2) A device for carrying film in a developing machine. (3) To adjust the exposure of a studio or *outside broadcast* video camera.

rack bars: The black lines between image frames on photographic film.

racking: (1) Another term for *framing.* (2) Rotating the *turret* on an old video or film camera to change the lens being used. (Hist.) (3) UK term for controlling the exposure and colour matching of studio video cameras from the *gallery.* See also *shading.*

rack-over: A movement to bring the *viewfinder (2)* of a film camera into line with its *objective lens* system.

radian: *SI unit* of angle, where 2π radians are equal to 360 degrees.

Radio Authority: UK government body which licenses and regulates all independent radio services. (Website: www.radioauthority.gov.uk).

Radiocommunications Agency (RA): The UK national radiocommunications regulatory body.

radio data service (RDS): Additional coded data transmitted within the bandwidth of FM audio *broadcast* for automatic tuning, station identification, continuous time check, etc.

radio frequency (RF): The frequency range of *electromagnetic waves (radio waves)* used for transmission, approximately between 20 kHz and 300 GHz. See **Appendix N.**

radio frequency interference (RFI): *Interference* introduced into electronic circuits by radio *transmitters* or other sources of electromagnetic radiation.

Radio-Keith-Orpheum (RKO): Founded in the USA in 1929, RKO is one of the oldest continuously operating film studios. (Website: www.rko.com).

radio link: *Point-to-point telecommunications* using radio waves, often in the *microwave* bands when it may also be called a *microwave link.*

radio microphone: A *microphone* connected to a small portable radio *transmitter* whose signal is picked up by a local *receiver*, thus avoiding trailing wires.

radio waves: See *radio frequency.*

RAID Advisory Board (RAB): An organization formed in 1992 which aims to standardize the terminology of RAID-related technologies. It publishes the *RAIDbook,* which aims to reflect 'the state of practice in storage systems'. (Website: www.raid-advisory.com).

RAIDbook: See *RAID Advisory Board.*

RAID array: A collection of magnetic disks connected and controlled in such a way as to provide some protection against loss of data due to a *disk* failure. See *redundant array of inexpensive disks* .

rails: Portable light-weight tracks for smooth operation of a *dolly* on *location.*

ramcorder: A *solid-state* digital video recorder based on *RAM* technology. (Hist.)

random access: The ability to select at will any defined point from a *slide magazine (2)*, programme or memory, as indicated by an address.

random access memory (RAM): Computer data memory which requires power to maintain its contents. Any locations can be accessed quickly and overwritten (compare *read only memory*).

random noise: *Noise*, or an electrical signal, that is random in its *amplitude, frequency* and *phase (1)* characteristics.

raster: The pattern of horizontal lines forming the image displayed in a television system.

raster graphics: Presenting graphical images as a *pixel*-based display on the *raster* of a *cathode ray tube* (compare *vector graphics*).

raster unit: The vertical distance between the midpoints of two adjacent *pixels.*

raw stock: Unexposed and undeveloped photographic film.

RCA connector: See *phono plug.*

read: To put information into a *computer* from a store; memory, *disk* or tape.

read only memory (ROM): Computer data memory that requires no power to maintain its contents. Data is permanently or semi-permanently stored. It is either impossible or difficult to write new data to the memory. It is a *non-volatile store* (compare *random access memory*).

real time: Keeping pace with events in the real world, as they are happening.

real-time programming: In *audio-visual* practice, programming a *tape-slide* show while the tape is running at its normal speed.

rear projection: The presentation of an image on a translucent *screen (1)* by a *projector* placed on the far side from the viewer .

reboot: To *reset* a *computer* system into its operating condition.

Rec. 601: See *ITU-R BT.601.*

Rec. 656: See *ITU-R BT.656.*

recce: Abbreviation for reconnaissance or reconnoitre. A survey of a location.

receiver (RX): A device that receives information, either in the form of an electrical signal or as an *electromagnetic wave* (e.g. in radio or television receivers).

recharging: Passing current through a *secondary cell* in the reverse direction to normal current flow to restore its capacity to provide electrical energy.

reciprocity failure: Divergence from the photographic *reciprocity law* which may occur at extremely high or low values of light intensity or time.

reciprocity law: In photography, the relation that constant *exposure* will be obtained if greater intensity of light is compensated by proportionally shorter exposure time and vice versa.

reclocker: An electronic circuit combining a *slicer* with a *phase locked loop*. It can remove *noise* and *jitter* from a digital signal.

Recording Industry Association of America (RIAA): The trade group that represents the US recording industry . Its mission is 'to foster a business and legal climate that supports and promotes our members ' creative and financial vitality '. Its members are the US record companies. (Website: www.riaa.com).

recovery time: The time taken for a device to return to normal after an overloading signal has been removed.

rectifier: See *diode*.

recursive: In computing, a *statement* or *subroutine* in which some steps make use of the whole *statement* or *subroutine*.

recursive filtering: A method of reducing video *noise* and random defects by the comparison of two or more adjacent *frames (2)*, non-repeating items being eliminated. The process must be adaptive to avoid *smear* in areas of motion and switch of f in those areas of the picture which are moving.

Red Book: The standard for *Compact Disc – Digital Audio (CD-DA)*.

red green blue (RGB): (1) The *primary colours* of an *additive colour* system in photographic or video colour reproduction. **(2)** A method of signal connection giving direct correspondence with the *primary colour* display areas on a picture monitor or *visual display unit*. **(3)** See *RGB video*.

Redhead: A portable mains-powered lighting unit, typically between 250 watts and 800 watts, with adjustable beam for spot or flood. (T rade name.)

reduction: The process of mixing signals from multi-track master tapes to produce a master tape for production. See also *mix down*.

reduction printing: Optically printing motion picture film to produce an image smaller than the original, usually on film of narrower gauge.

redundancy: (1) In video data *compression (2)*, repetitive information which need not be encoded to allow a good approximation of the original pictures to be reconstructed by the receiving device. **(2)** Duplication of equipment or systems to ensure continued operation if one device or system fails.

redundant array of inexpensive disks (RAID): An interconnection of magnetic or optical *disks* for storing data which gives protection for the data in the event of a single *disk* failure. *Parity*-based *error correction codes* are stored in the array along with, but separate from, the data to which they refer. They allow the data on a corrupted or faulty *disk* to be reconstructed completely and restored to a new replacement *disk*. See also *RAID array* and *RAID Advisory Board.*

Reed-Solomon coding: A byte-based *forward error correction* scheme. In its application in *MPEG-2 transport streams*, 16 bytes of error correcting *parity* codes are added to each 188 bytes of data. This allows up to 8 bytes of error in a 204 byte packet to be fully corrected and is denoted by RS (204, 188, 8).

reel: (1) See *spool.* **(2)** A roll of film, the unit in which a programme or part of a programme is usually handled, either as the assembled negative or corresponding positive print.

reel band: A paper strip securing a tightly wound roll of motion picture print.

reel-to-reel: A film or *tape path* with separate feed and take-up *spools* in a camera, *projector* or *magnetic tape* recorder/reproducer, in contrast to an enclosed *cassette* or *cartridge.*

reflector: A white or silver sheet used in location shooting to bounce natural light to illuminate areas in shadow .

reflex camera: A still photographic or film camera in which the image in the *viewfinder (2)* is obtained through the main *objective lens,* so that *parallax* error is eliminated. See also *through the lens.*

reflex projection: See *front axial projection.*

refraction: The deflection of a beam of radiation, such as light, when passing from one medium to another .

refractive index: A factor expressing the *refraction* caused by a medium. For light it is the ratio of the speed of light in a vacuum to the speed of light in a medium. This can also be expressed as the ratio of the sine of the angle of incidence to the sine of the angle of *refraction* for a ray of light passing from a vacuum into a medium, using Snell 's law, defined by Willebrord Van Roijen Snell, a seventeenth century Dutch astronomer and mathematician. It is called *index of r efraction* in the USA. See **Figure R.1**.

refresh: (1) To regularly rewrite data into a *volatile* digital memory so that its contents are not lost. **(2)** In *computer graphics,* to update the display on the picture monitor .

refresh rate: In computing, the number of times each second the image stored on a graphics card is redrawn on the computer display screen.

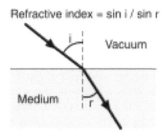

Figure R.1 Snell's law of refraction

region code: A digital code on a *DVD*-Video *disc* restricting playback to players with the appropriate region number . See **Appendix Y**.

register: (1) A short term store for digital information in a *computer central processing unit*. (2) To place in exact alignment or position.

register bar: A locating device consisting of a metal strip with two or more upstanding pegs, used to ensure accurate placing of material to be photographed or copied. The drawing or *cel* is edge perforated to match the pins.

register mount: A *slide mount* containing locating pegs for precise location of the film by means of its perforations.

register pegs: Pegs on a *light box* or *rostrum (2)*, on which *cels* can be placed and kept in register .

register pins: Locating pegs accurately shaped to fit into the edge perforations of film, paper or transparent sheet. In a camera, register pins precisely locate the film by means of its perforations.

registration: (1) The precise location of successive images in a sequence so that unintentional movement is not introduced. **(2)** The correct positioning of three-colour component images to be coincident without overlapping or fringing. This operation was necessary with three- *tube (2)* colour video cameras but is not required or even possible with video cameras using *CCD imagers* since the imagers are permanently registered and sealed in position during manufacture (compare *convergence (1)*).

registration slides: See *line-up slides*.

release print: A motion picture positive print of picture and sound made for general distribution and exhibition. See also *inter-negative*.

remanence: The residual magnetism of a ferromagnetic substance when, after being magnetized to saturation, the magnetizing field is removed.

rem-jet: A back coating applied to photographic film to protect against *halation*, static and scratching. It is removed during processing.

remote: See *outside broadcast.*

remote control: Operating a device from a distance, by cable, radio, *infrared* or other means.

repeater: A *transmitter* which re-broadcasts the signal it receives from another *transmitter.*

reporter: A commentator appearing in front of camera.

Request For Comments document (RFC): A series of publications from the *Internet Architecture Board* on *Internet* technical matters.

Research into Advanced Communications for Eur ope (RACE): A series of European Community research programmes that ran between 1992 and 1995. (Hist.)

reseau: A network or mosaic of small *elements (5),* as used in some forms of colour reproduction process.

reset: To restore a device to defined conditions, usually corresponding to those present at start up.

residual current device (RCD): An electrical safety device that measures the difference in current flowing between the supply wires. The supply is rapidly disconnected (a typical rating is 40 ms) if a dif ference above a certain threshold is detected (a rating of 30 mA is typical).

resolution: The ability of an optical or video system to reproduce fine detail. A measure of the 'sharpness' of a picture. Usually measured in *pixels* in the horizontal direction and lines or *pixels* vertically. Often the horizontal and vertical resolutions are dif ferent.

resolving power: The *resolution* of a system expressed in numerical terms.

resonance: A sympathetic oscillation of a mechanical or electrical system when stimulated at its own natural frequency .

retake: The re-photography of a scene.

reticle: A ground glass *screen (1)* in a film camera *viewfinder (2)* engraved with the exact size and position of the camera gate, centre cross lines, etc. See *graticule.*

reticulation: A pattern of small cracks and wrinkles on a photographic emulsion surface, caused by incorrect *processing.*

retina: The internal surface of the human eye behind the *lens (3)* which contains millions of light-sensitive cells. See also *rods, cones, human eye* and **Figure H.6.**

retrace: See *flyback.*

re-transmit (RX): To send again.

retrieve: In computing, to read or recover stored data.

retrofocus lens: An *objective lens* where the secondary *principal point* is behind the lens. It corresponds to the wide-angle end of a zoom lens with a short *focal length* (compare *telephoto lens*). See **Figure P.10.**

return: See *carriage return.*

return loss: A measure of the amount of incident power that is reflected toward the opposite end of a cable from a *termination* or other discontinuity in the *characteristic impedance* of the cable.

return to zero (RTZ or RZ): A serial data stream in which the logic level returns to zero after each one during the transmission of consecutive data ones (compare *non-return to zero*).

reveal: See *conceal/reveal*.

reverberation: The sustaining of sound within a structure such as a room or building due to short term multiple reflections (compare *echo (2)*).

reverberation time: The time taken for a sound to die away by 60 *decibels* in *amplitude* after the direct sound has ceased (60 *decibels* is one millionth of its original value). See also *RT60*.

reversal film: A photographic film which after processing yields an image whose tonal distribution matches that to which it was exposed. For example, a camera film giving a positive image, or a duplicating *stock* giving a negative image from a negative.

reverse: To step back one *slide* with a *slide projector* or back one image with a *dissolve pair*.

reverse shot: A shot of a performer's face from the approximate viewpoint of a second performer or interviewer. See also *over the shoulder shot*.

rewind: To *transport* a film or tape in the reverse direction, back onto its original hub or *spool*.

RF cues: In motion picture printing, *cues* applied to the edge of the negative in the form of small metal patches that are sensed by a *radio frequency* detector.

RGB video: *Component video* with separate red, green and blue video signals.

RIAA curve: The universally adopted *Recording Industries Association of America* recommended audio reproduction/ *equalization* characteristic.

ribbon microphone: A sensitive, high quality *bi-directional microphone* operating by currents induced in a thin metal ribbon mounted in a magnetic field.

rifle microphone: A highly *directional microphone* which can be aimed at its sound source. Also called a *shotgun microphone*. It is tubular with slots along its length, which tend to attenuate sounds arriving from any other direction than where it is pointed.

rigging: The setting up of lights, *loudspeakers* or *projectors* on a *set*, stage or venue.

right-hand circular polarization (RHCP): See *circular polarization*.

rim light: Lighting the edges of the subject, usually from behind.

ring: The central contact on a three-circuit *jack* plug. See **Figure P.8**.

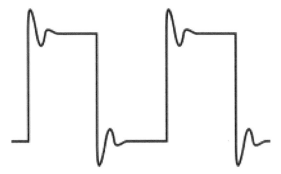

Figure R.2 Ringing

ringing: A damped oscillatory response at the beginning or end of a *pulse* signal. In a video display , it appears as closely spaced light and dark bands of decreasing intensity at the edge of an object where a marked change of *brightness* occurs. See **Figure R.2**.

ring modulator: A device requiring two inputs to produce an output consisting of sum and difference frequencies between the signals. Used in electronic music synthesizers.

ring network: A *network topology* where the *nodes (2)* are connected together in a closed loop and there is no central *computer* to control the system. *Token ring* is an example of a *network* system using this *topology.*

ripple: (1) An unwanted *AC* signal superimposed on a *DC* supply. **(2)** A rapid succession of cuts or *dissolves* across a series of *screen (1)* areas of a *multi-screen* presentation. **(3)** A *mix* effect in which the outgoing and in-coming scenes are distorted behind rippling lines.

rise time: The time taken for the leading edge of an electrical pulse to rise between two fixed levels, usually from 10 per cent to 90 per cent of its maximum *amplitude* (compare *fall time*).

risers: The vertical steps of a *Fresnel lens*.

RJ-45: The standard plug for *Ethernet* and other systems (e.g. *Integrated Services Digital Network*) based on *twisted pair* cabling. Also called after its definition in standard *ISO 8877*. See **Figure R.3** and **Table R.1**.

roam: In a *computer graphic* display, to move the *window* through which a part of the image is accessed and observed.

rock and roll: The ability to move a *videotape* recording or *synchronized* picture and sound films backwards and forwards at varying speeds, to decide an *edit point* or for convenience in *dubbing* and *mixing*.

rod aerial: A directional aerial used at *microwave* frequencies.

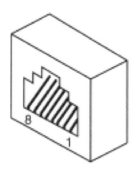

Figure R.3   RJ-45 connector

Table R.1   RJ-45 connector pinouts for various applications

Application	Pins			
	1 & 2	*3 & 6*	*4 & 5*	*7 & 8*
Analogue telephone			Tx/Rx	
ISDN-2 (basic rate interface)		Data Rx	Data Tx	Power (Optional)
Ethernet (10BaseT)	Data Tx	Data Rx		
Token ring		Data Tx	Data Rx	
FDDI	Data Tx			Data Rx
Ethernet (100BaseT 4-pair)	Data	Data	Data	Data
Ethernet (100BaseT 2-pair)	Data	Data		
ATM LANs	Data Tx			Data Rx

rods: The cells in the *retina* of the human eye which are sensitive to movement and low levels of light (compare *cones*). See also *human eye* and **Figure H.6**.

roll: (1) To start one or more film cameras, audio or *videotape recorders*. **(2)** A *reel (2)* of photographic film.

roll feed: An attachment to an *overhead projector* to feed a roll of transparent material across the *platen*.

rolling title: A series of *titles* or *captions* moving upwards across the picture area. Also called *creeping title*.

root mean squar e (RMS): A factor giving the ef fective value of an alternating current or voltage, taking its waveform into account. For a current, it results in the same heating ef fect as a direct current.

roping: Film damage in the form of continuous *sprocket*-tooth indentations, caused by *run-off.*

rostrum: (1) A platform raising actors and scenery, often to camera height. **(2)** A camera stand designed to illuminate and precisely hold the artwork to be filmed in position.

rostrum camera: A fixed film or video camera, mounted vertically for shooting complicated graphics and *animation.*

rotary erase head: See *flying erase head.*

rotary printer: A continuous *contact printing* machine in which both films are carried on a rotating *sprocket* at the time of *exposure.*

rotary wipe: See *clock wipe* and **Figure W.1.**

rotations: Effects obtained by using the *turntable (2)* on an *animation stand.*

Rotoscope: An instrument projecting a film frame by frame on to the table of an *animation stand,* for the preparation of *animated* drawings or as a *background (4)* for *animation cels.* (Trade name.)

rough animation: The first tentative *animation* done as a test.

rough cut: In *editing,* the first assembly of shots in their intended script order.

router: (1) A sophisticated data *network* device which can choose paths for data on the basis of *protocols* and *addresses (3).* **(2)** In a radio or television studio centre, a *matrix* which passes signals from one location to another within the centre.

routine: A name for a short computer *program* which may be used several times.

routing matrix: See *matrix.*

Royal Photographic Society (RPS): This UK Society was formed as The Photographic Society in 1853 with Queen Victoria and Prince Albert as patrons. It was granted its royal title in 1894. The Society's mission has always been 'to promote the Art and Science of Photography'. (Website: www.rps.org).

Royal Television Society (R TS): The only UK Society exclusively devoted to television which began as the Television Society in 1927, nine years before the first public service broadcast from Alexandra Palace. It was granted its royal title in 1966 and represents over 4000 members from across the broadcasting industry . (Website: www.rts.org.uk).

ROYGBIV: Red, orange, yellow , green, blue, indigo and violet. A mnemonic for the order of colours in the *visible spectrum,* i.e. outside this range are *infra-red* and *ultraviolet.* See also *VIBGYOR.*

RS standard: Recommended Standards of the *Electronic Industries Association*, including *RS-232*, *RS-422* and *RS-423*.

RS-232: A serial data interface standard, used for the serial ports on PCs. It defines the electrical characteristics and pin allocations in the 25-way connector.

RS-422: A balanced version of *RS-232*, used for *VTR* remote control systems and the serial ports on Macintosh computers. It is a dif ferential 5 *volt* system and therefore less prone to interference than *RS-232*.

RS-423: As *RS-422* but with a high *impedance* state to allow more than one sending circuit to be active.

RT60: The *reverberation time* for a 60 *decibel* drop in *amplitude* (60 *decibels* is one millionth of its original value).

rubber numbers: Numbers and other coding applied to processed *rush prints* and sound records for identification during film *editing*.

rubbery: Describes the look of pictures and sound which drift in and out of lip sync, often as a result of *post syncing*.

rumble: A low frequency vibration of mechanical origin which becomes superimposed on the wanted signal. Often associated with *audio disc turntables (1)*.

run: (1) The instruction to commence execution of a computer *program*, usually from the beginning. **(2)** The length of time for which a theatrical production is put on. **(3)** The number of episodes in a radio or television series.

runner: The most junior member of a production team, whose role might include collecting tapes, looking after actors and general administrative functions.

run off: Displacement of film so that it passes over the teeth of a sprocket, with resultant damage.

run out: (1) A length of blank film *stock* at the end of each *reel* to protect the film from damage. **(2)** The section of black film immediately following the last picture *frame (1)* in the end section of an *Academy leader*.

run through: Rehearsal.

run time: Taking place when a *computer* is operating. (As opposed to when a *program* is entered.)

run up: The length of film or tape that has to run through a *projector* or audio or *videotape recorder* before it is operating smoothly at normal speed.

rushes: (1) See *rush prints*. **(2)** Unedited *videotape* from *ENG*.

rush prints: In motion picture practice, the first prints made from newly processed picture or sound negative shot each day during production, used to check content and quality . Also called *dailies* or *rushes*.

S

safe action area: An area in the centre of the *active picture* where action is positioned to ensure that none of it will be cut of f by the *overscan* on a domestic television receiver . It corresponds to 90 per cent of the transmitted *active picture* area of a video image. See also *safe title area* and **Appendix B**.

safe area: The area within the *active picture* which is considered to be always visible on a domestic television receiver . See also *safe action area, safe title area* and **Appendix B**.

safe area generator (SAG): Equipment that displays lines on a monitor corresponding to *safe action ar ea, safe title ar ea* and perhaps other markings.

safelight: A source of visible light in a photographic darkroom whose colour and low intensity allow unprocessed light-sensitive materials to be handled without danger of unwanted *exposure.*

safe title area: An area in the centre of the *active picture* where a *title* or other graphic is positioned to ensure that none of it will be cut off by the *overscan* on a domestic television receiver. It corresponds to 80 per cent of the transmitted *active picture* area of a video image. See also *safe action area* and **Appendix B**.

safety base: The almost non-flammable *acetate (1)* or triacetate film base that has replaced the highly flammable nitrate base formerly used.

sample: A digitized signal is composed of samples, each of which represents the digital value of the signal at a moment in time.

sample and hold: An electronic circuit which measures the instantaneous *amplitude* of an electrical *waveform* and stores (holds) the value for a period of time. Commonly used in *analogue-to-digital converters.*

sample print: See *answer print.*

sampling: The process of measuring the *amplitude* of an *analogue signal* at regular intervals in time. See **Figure A.2** and **Figure Q.1**.

sampling rate: The frequency at which an analogue input signal is *sampled* in an *analogue-to-digital converter.* See also *Nyquist frequency.*

sandcastle pulse: In electronics, a *waveform* able to convey varied timing information by slicing it at dif ferent levels. (Hist.)

sandwiching: Mounting two or more pieces of film in a single *slide mount* to create effects.

satellite: (1) A sub-station of a main location. **(2)** An unmanned space vehicle, usually orbiting the Earth and providing communications, astronomical or other scientific information or functions. See also *telecommunications satellite, downlink, uplink, LEO, MEO* and *GEO.*

Figure S.1   Sawtooth signal

Satellite Broadcasting and Communications Association (SBCA): The US national trade organization representing all segments of the *satellite (2)* industry, founded in 1986. It is committed to expanding the use of *satellite (2)* technology for the broadcast delivery of entertainment, news, information and educational programming. (W ebsite: www.sbca.com).

satellite master antenna television (SMA TV): Cable or *microwave* distribution of *satellite (2)* broadcasts to individual *viewers (2)* from a central *aerial*.

satellite news gathering (SNG): Gathering news material for radio or television news broadcasts using *satellite (2)* communication systems.

satellite newsgathering vehicle (SNV): A self-contained mobile unit for *SNG*.

satphone: A *satellite (2)* telephone, usually used on one of the *Inmarsat* systems.

saturation: (1) The state when a magnetic material is fully magnetized. **(2)** In colour reproduction, the spectral purity (intensity) of a colour .

sawtooth: An electrical *waveform* which has a slow change in *amplitude* followed by a rapid return or vice versa. It is the basic *waveform* used in the *scanning* circuits of *cathode ray tube* based display devices. See **Figure S.1**.

sawtooth test signal: A video test signal *waveform* which rises or falls evenly between set limits. See **Figure S.2**.

scan line: A horizontal line on a television picture display device.

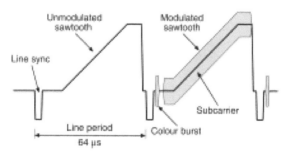

Figure S.2   Sawtooth test signal waveform

scanner: **(1)** The spinning video *head drum* assembly in a *videotape recorder.* **(2)** Abbreviation for *caption scanner.* **(3)** A mobile video production vehicle. **(4)** A computer *peripheral* for *scanning* photographic images.

scanning: The process by which an area is systematically imaged line-by-line, particularly in television, e.g. where an electron beam is used in a *cathode ray tube.*

scanning yoke: See *yoke.*

SCART connector: *Syndicat des Constructeurs d'Appar eils Radio-récepteurs et T éléviseurs.* This connector is the analogue video interconnection standard for consumer equipment such as television *receivers, video cassette recorders, videodisc* players, *computers,* cable and *satellite (2)* receiving systems. It is also called a *Peritel* or *Euroconnector.* See **Figure S.3** and **Table S.1.**

scenario: See *script.*

Scene Sync: A system in *chromakey* special effects which permits movement of the foreground camera synchronized with the mask camera. (Trade name.)

SCH phase: In *composite video (2)* systems, the phase relationship between the colour *subcarrier* and the *leading edge* of *line sync,* defined in *PAL* on line 1 of field 1.

scientific notation: Representing a lar ge or small number as a number between 0 and 10 and a power of ten, e.g. 1 252 000 as 1.252×10^{6}.

Schmidt optical system: A large *aperture (1)* optical projection system or telescope having a spherical concave mirror in combination with an *aspheric* correction lens. Named after the Estonian-born German optician Bernhard Voldemar Schmidt. See **Figure S.4.**

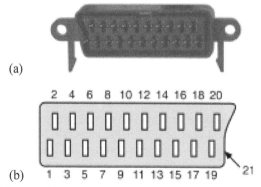

(a)

(b)

Figure S.3 SCART connector

Table S.1   SCART connector pinout

Pin	Signal name	Pin	Signal name
1	Audio Out Right	11	RGB Green In
2	Audio In Right	12	Data 1: Data Out
3	Audio Out Left + Mono	13	RGB Red Ground
4	Audio Ground	14	Data Ground
5	RGB Blue Ground	15	RGB Red In / Chrominance (C)
6	Audio In Left + Mono	16	Blanking Signal
7	RGB Blue In	17	Composite Video Ground
8	Audio / RGB Switch	18	Blanking Signal Ground
9	RGB Green Ground	19	Composite Video Out
10	Data 2: Clockpulse Out	20	Composite Video In / Luminance (Y)
		21	Connector Shell – Chassis Ground

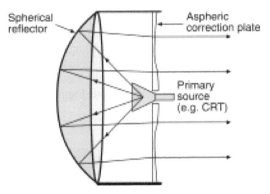

Figure S.4   Schmidt optical system

Schmitt trigger: An electronic circuit whose output increases rapidly to a maximum when the input rises above a certain threshold level and decreases rapidly almost to zero when the input falls below another lower threshold level. This hysteresis means that a slowly changing input gives rise to a sharply defined and unambiguous output. Named after the American biophysicist and electronics engineer Otto H. Schmitt.

scoop: A *luminaire* in the form of a spun aluminium bowl used as a soft source for general *fill* lighting.

scope: (1) An *oscilloscope.* (Colloq.) **(2)** A *vectorscope.* (Colloq.) **(3)** *Cinemascope.* (Colloq.)

Scotchlite: A highly reflective beaded *screen (1)* material for *reflex projection.* (Trade name.)

scotopic vision: The response of the human eye to electromagnetic radiation in low light conditions. The peak in response shifts to about 500 nm so that in twilight objects appear bluer. In very faint light colour vision is lost altogether as the *rods* dominate (compare *photopic curve*). See also *Perkinje effect.*

Scottish Media Gr oup (SMG): The parent company of Scottish Television and Grampian Television in the UK. (W ebsite: www.smg.com).

scrambling: The continuous modification of a transmitted signal so that it can only be correctly received by use of an authorized descrambler .

scrape flutter: A fault in magnetic recording resulting from alternate sticking and slipping motion of the recording tape producing *flutter.*

scratchback: In animation, a technique in which part of the artwork, painted on a *cel,* is removed frame by frame under the camera while the camera runs backwards.

scratch pad: A memory area in a computer system used for holding temporary data.

screen: (1) The surface on which a picture image is presented. **(2)** A conducting surface preventing external electrical or magnetic *interference* with a system. **(3)** A braided or lapped conducting layer around an inner signal conductor in a *coaxial cable.* **(4)** To project a slide or film onto a surface.

screened cable: A central conductor, or a number of conductors, insulated from and within a *screen (3)* formed by copper braiding or lapping or foil.

screen dump: A copy of the screen display on a *computer* which is sent to a printer or *disk.*

scrim: (1) A gauze used to dif fuse light. See also *butterflies.* **(2)** A metal gauze used to provide protection for bare light bulbs.

script: (1) The detailed scene-by-scene instructions for a film or television production, including a description of the setting and action with dialogue and camera directions. Also called a *scenario.* **(2)** A similar treatment for a *tape-slide* or multivision production. When the script also shows full details of the visuals, it is termed a *story board.*

scrix: A video ef fect where the *frame (2)* is divided into a number of rectangles which reduce and increase in size and are displayed in a different order on the *screen (1).*

scroll: (1) The part of a vinyl *audio disc* surface where the pitch of the grooves has been increased to separate bands. **(2)** A roll of transparent film for use on an *overhead projector*. **(3)** A video *transition* similar to *push-on, push-off,* but with the picture displaced vertically . **(4)** In a computer display , the addition of a bottom line which displaces all the others upwards. With *spreadsheets,* a similar effect in any direction.

scrolling: Adding a new line of information and moving the others vertically.

seamless masks: See *soft-edged masks.*

search: An ability to move rapidly backwards and forwards to other parts of the programme without loss of sound-picture synchronization.

search time: The time taken to complete a search and find the required item or part of the programme.

second: *SI unit* of time.

secondary: A *transformer* winding other than the *primary.*

secondary battery: An assembly of *secondary cells.*

secondary cell: An electrochemical *cell* which generates electrical energy by chemical changes in its electrodes. The process can be reversed by *recharging.*

secondary colour: In light, a colour produced when two *primary colours* are mixed together.

segment: In *computer graphics,* a number of display items which can be manipulated as a single unit.

sector: A division of a *track (3)* on a *hard disk drive* or *floppy disk.* See also **Figure H.2.**

segue: An indication that one section of music is to be played immediately after another one. The changeover may be a straight cut or a *cross-fade.*

Selectavision: A grooved capacitance *videodisc* system. (Trade name.) (Hist.)

self-blimped: Describing a motion picture camera whose operating *noise* level is so low that no additional sound-proof enclosure is necessary .

self clocking: A data transmission system using a single channel to carry the data and the clocking information needed to decode it. See also *channel coding.*

self key: A *keying* effect where one video signal is both the *key signal* and the *fill (1).*

sell-through: Distributing *videograms* by direct retail sale to the public rather than by rental.

Selsyn: An electrical *servo system* using three-phase motors to *synchronize* a *telecine* machine with a separate audio replay machine. (Trade name.) (Hist.)

senior: A 5 kW studio spotlight with a *Fresnel lens*. (Colloq.)

sensitometry: The scientific study and measurement of the ef fect of light on photographic materials, especially the relation between *exposure* and the resultant *density* after processing.

separation: Acoustic isolation between instruments essential in multi-track recording to give control over relative balance between instruments.

separations: See *colour separations.*

sepmag: A magnetic sound record, on tape or perforated film, separate from the associated picture film.

sequential access: A mode of data retrieval where each byte of data is recovered in the order in which it was written to the disk.

Séquentiel Couleur à Mémoire (SECAM): Sequential colour with memory. The analogue colour television system proposed by Henri de France in 1958 and further developed by the Compagnie Fran çaise de Télévision in Paris and used in France and in most of Eastern Europe (see **Appendix I** for other countries). It uses 625 lines per *frame (2)* and 50 *fields* per second. The two *colour difference signals* are transmitted on alternate lines as *FM* signals.

sequential scanning: See *progressive scanning.*

serial communication: A data communication scheme using a single pair of connections where data is sent one bit at a time.

Serial Data Transport Interface (SDTI): A data stream specification for transporting packetized data within a television studio or production centre environment, defined in standard SMPTE 305M (see **Appendix U**). The data packets and synchronizing signals are compatible with the *Serial Digital Interface (SDI)* .

Serial Digital Data Interface (SDDI): A proprietary interconnection standard developed by Sony as an extension to *SDI* for carrying data signals using *coaxial cable* and *BNC* connectors. It is electrically compatible with *SDI.*

Serial Digital Interface (SDI): An interconnection standard developed by Sony for digital *broadcast* equipment using *coaxial cable* and *BNC* connectors. It subsequently became part of the *ITU-R BT.656* standard and is also defined in standard SMPTE 259M. An optical fibre version is defined in SMPTE 297M. See **Appendix U.**

serializer: An electronic device which converts a parallel data stream to a serial one (compare *deserializer*).

server: A storage system for data files which also provides services to (serves) users connected to it (clients) on a *network.* Hence this arrangement is called a *client-server network* or server based network (compare *peer-to-peer network*). See also *e-mail server, file server, print server* and *video server.*

service: In *Digital Video Broadcasting*, a service is a broadcast television or radio channel, e.g. BBC 1 or Capital Radio in the UK. See also *program (2)*.

servo system: A control system involving *feedback*, which operates to correct any discrepancy between the required output and a reference input, often using a sensor.

set: In a film or television production, the studio floor area where the action is to be played, including the scenery.

set level: In multivision control systems, a continuous tone at an exact level which is used to adjust recording level.

set light: See *background light*.

set-top box (STB): A stand-alone unit connected to a television *receiver* for decoding *satellite (2)* broadcast or *cable television* signals. See also *Integrated Receiver Decoder (IRD)*.

set up: (1) The complete system. **(2)** In motion picture production, the arrangement of setting, actors, lights, *microphones* and cameras ready to record a scene. Also, a specific camera position. **(3)** To get into a state of readiness to operate. **(4)** In a video system, to differentiate between zero level or true black, and the actual black level of the reproduced picture (UK). **(5)** The US term for *pedestal (2)*.

shade: A colour mixed with black, e.g. brown is a shade of yellow.

shading: US term for adjusting the picture quality of television studio cameras. See also *racking (3)*.

shadowboard: A black board suspended below the camera on an *animation stand*. It has a hole through which the lens protrudes to prevent reflection of the camera in the *platen* glass.

shadow key: A *chromakey* effect which retains the shadows cast by foreground objects.

shadowmask tube: A *cathode ray tube* for colour television displays having a mosaic of *RGB* phosphors at which three electron beams are directed through a perforated metal screen (the shadowmask) (compare *aperture grille*). See also *delta gun tube* (**Figure D.2**), *in-line tube* (**Figure I.1**), and *slot mask tube*.

sharp: In focus, optically and electronically.

sharpness: See *resolution*.

shash: The noisy picture and hissy sound from an untuned television receiver.

shedding: Losing particles of oxide from a magnetic tape often due to faulty manufacture or poor storage conditions.

shielded twisted pair (STP): *Twisted pair* cable with an earthed *screen (3)* to increase its immunity to *interference*. It is used in high-capacity data networks.

shielding: (1) In *computer graphics*, defining an opaque *window* in which to display a message. **(2)** The *screen (3)* on a *coaxial cable*.

shipping reel: A heavy-duty 2000ft *spool* used for film *release print* distribution in the United States.

shoot: To operate a camera.

short end: The portion of a roll of film or *magnetic tape* remaining after the main part has been used.

shorty: See *baby legs*.

shot: A scene photographed or recorded as one continuous action. See also *take (1)*.

shot box: A box mounted on a video camera which has several buttons to activate preset automatic zoom speeds and zoom positions.

shotgun microphone: See *rifle microphone*.

show copy: A selected copy of a completed programme, video, *tape-slide*, multivision or film, intended for presentation to an audience. For a motion picture production it is often termed a *show print*.

Showscan: A high-quality cinema presentation system using 35mm film projected at 60 *frames per second*. (Trade name.)

shufflecasting: Repeated transmission of a programme on different days.

shutter: (1) In a photographic camera, a mechanism which is opened and closed in a preset time to effect the *exposure* by controlling the amount of light falling on the film. **(2)** In motion picture equipment, a rotating blade which interrupts the light in a camera or *projector* while the film *pull-down* occurs. **(3)** In *slide* projection, a device to cut off the light while the *slide* is being changed or if there is no *slide* in position. See also *snap*. **(4)** In a video camera, an electronic shutter is used to alter the timing and amount of light from the viewed scene falling on the *CCD imagers*. It can also be used to *synchronize* the camera with a scanned display which is in shot to eliminate visible screen flicker.

shuttle search: The ability to play back a film, audio tape or *videotape* in either direction and reproduce the picture over a wide range of speeds.

sideband: A range of frequencies above or below a *carrier* frequency generated by *modulation*.

sidereal day: The rotational period of the Earth, about 23 hours, 56 minutes and 4.1 seconds. It is less than 24 hours since the Earth moves in its orbit around the sun a little each day, effectively shortening the sidereal day by 1/365 of a day each day(!)

siemens (S): *SI unit* of electrical conductivity, the reciprocal of an *ohm*, equivalent to *mho*. Named after four nineteenth century German-born brothers, Ernst Werner, Karl Wilhelm, Friedrich and Karl von Siemens.

signal-to-noise ratio (SNR or S/N): The ratio in *decibels* of the received *signal* power to the *noise* power in a given bandwidth for an analogue system. In video, a S/N ratio of between 54 dB and 56 dB is considered full broadcast quality.

silicon controlled rectifier (SCR): See *thyristor*.

simple PAL: A PAL decoding system where the colour is displayed directly with any phase errors visible as alternating pairs of lines with incorrect *hue* (compare *delay PAL*).

simplex: A mode of data transmission in a single direction only .

simulcast: Simultaneous *broadcast* of a programme on two dif ferent channels, for example, on both radio and television, on both AM and FM radio, on both analogue and digital television broadcast services, on television services with both standard and high *definition* pictures or on television services with 4:3 and 16:9 *aspect ratio* images.

Simul-sync: A method of avoiding the time delay caused by the distance between the recording and replay *heads (2)* on a tape recorder by replaying from the record *head (2)*. (Trade name.)

sine wave: A *waveform* showing variation in a simple harmonic manner .

single channel per carrier (SCPC): Sending only one video service on a *satellite (2) transponder*.

single frame: (1) One individual image in a strip of images. **(2)** To expose or project one picture at a time at a comparatively slow rate.

single frequency network (SFN): An optional configuration for a digital broadcast network using *OFDM* where all transmitters can send the same digital broadcast at the same frequency and not interfere with each other.

single mode fibre: *Optical fibre* with a *core (2)* diameter of typically less than 10 μm in which only one mode of light propagation is possible (compare *multimode fibre*). Also called *monomode fibre*.

single shot: (1) Exposing film, one *frame (1)* at a time, as in *animation* work. **(2)** A *trigger (4)* mode for an *oscilloscope* that allows it to display an event in a signal which only occurs once (also called a *transient* event).

single standard: A video monitor, *videotape recorder* or other equipment which only operates with one specific system.

single system: A method of motion picture production in which both picture and sound are simultaneously recorded on the same strip of film.

sit: Depressing a video signal towards black and therefore darkening the picture.

SI unit: *Système International d'Unités*. The system where a physical unit is derived from the basic units of the *metre, kilogram, second, ampere, kelvin, candela* and *mole*.

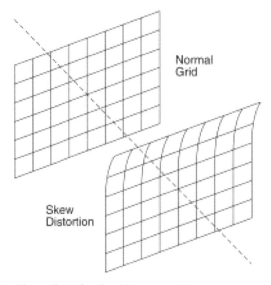

Normal
Grid

Skew
Distortion

Figure S.5   Skew distortion in videotape replay

skew: (1) A video picture distortion in which the verticals are not at right
angles to the horizontals. It can also be artistically introduced as an
effect. **(2)** The non-simultaneous arrival of data on several data lines in
a parallel interconnection. **(3)** Fine adjustment of a *polarizer* on a
motorized *dish aerial* covering a large portion of the *geostationary arc*.
(4) In videotape recording, a curved distortion at the top of the replayed
picture due to mechanical errors in the tape motion or incorrect tape
tension. See **Figure S.5**.

Skillset: The UK national training or ganization for broadcast, film, video
and multimedia. (Website: www.skillset.org)

skin effect: The tendency for *high frequency* signals to travel near the
surface of conductors.

skip frame: In *cinematography*, printing only *frames (1)* selected at
regular intervals from the original record to produce the ef   fect of
speeding up action.

skivings: Fine hair-like slivers of film stripped from its edge by an
obstruction during running.

skylift: A lorry mounted hydraulic platform able to reach up to 66 metres
from the ground.

skypan: A large *luminaire* with a matt white reflector for lighting back
drops.

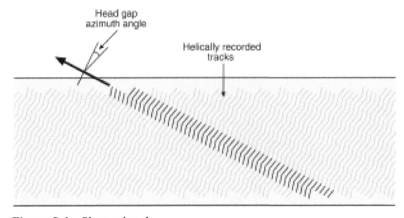

Figure S.6   Slant azimuth

slant azimuth: In magnetic recording, having the *head gap* at an angle other than 90 degrees to the direction of tape movement. Different slant azimuths on adjacent tracks reduces *crosstalk* without the use of *guard bands*. See **Figure S.6**.

slant range: The length of the *line of sight* path between a communications *satellite (2)* and an associated *earth station*.

slave: A device designed to function only as ordered by a 'master' unit, e.g. one *videotape recorder* that is controlled by another.

slate: A chart or board with information regarding the production, its title, scene and *take (1)* numbers, etc. reproduced at the beginning or end of each *take (1)* and photographed on film or recorded on tape. Also called a *number board*.

slewing: Synchronizing two or more *videotape recorders* involved in an *edit*.

slew rate: The rate of change of a signal.

slicer: An electronic circuit whose output level depends on the input being above or below a threshold level. It is used to remove *noise* from a digital signal. See also *reclocker*.

slide: The complete assembly of a still picture *transparency* in a mount, which may also contain a mask.

slide mount: A metal or plastic *frame (8)*, often glazed, for holding a picture image on film.

slide-tape: See *tape-slide*.

slider: A form of *potentiometer* with a linear rather than rotary control.

sliding frequency: See *gliding frequency*.

slit: (1) In photographic sound, a narrow slot through which the film is exposed in recording and scanned in reproduction. **(2)** To cut a wide roll of *magnetic tape* into the width needed for the *cassette* or *reel* format, e.g. 12.5 mm (½ inch) for *VHS*.

slit loss: In photographic sound, the reduction in *amplitude* at higher frequencies which results from the finite height of the *slit (1)* in recorder or reproducer.

slo-mo: Abbreviation for *slow motion*. (Colloq.)

slope: The steepness of a *filter (3)* output response usually expressed in *dB/ octave*.

slot: In a *satellite (2)* system, the orbital position in the *geostationary Earth orbit*, allocated by the *ITU*, in which a *satellite (2)* is located.

slot loading: Describes a mechanism where tape or film can be *loaded (4)* laterally instead of by *lacing* or threading.

slot mask tube: A *shadowmask tube* in which the mask and phosphors are in vertical stripes rather than dots. See *in-line tube* and **Figure I.1**.

slow motion: Photography by a motion picture camera with the film running faster than normal, so that when the result is projected at normal speed, the action appears to be slowed down. The same effect is achieved in *videotape* recording by either using *dynamic tracking* or by electronically *interpolating frames (2)* in between those actually recorded or by recording on a special high-speed *videotape recorder*.

slow scan: *Scanning* and transmitting video at a lower *frame rate* than normal. It is sometimes used in *video conferencing* to reduce the required processing and signal *bandwidth* for transmission.

small computer system interface (SCSI): A means by which up to 8 *computers* and/or *peripherals* such as optical or magnetic disk drives or tape drives can be connected together to allow data and control signals to be conveyed rapidly between them on a 50-way cable.

smart-card: A plastic card, similar to a credit card, which has a microchip and memory instead of a magnetic strip. It can typically store about the same amount of data as three pages of typewritten data.

smear: A video picture defect in which objects appear to blur in the direction of motion when they move.

smidgen: A small technical adjustment. See also *gnats*. (Colloq.)

snap: A very rapid picture change ef fect achieved using *shutters (3)* in *slide projectors*.

snoot: A cylindrical or conical hood to reduce the width of the beam from a light source. (Colloq.)

snow: Random *noise* or *interference* appearing in a video picture as white or black specks.

snubber: An electronic circuit which suppresses high frequencies.

Société Européene des Satellites (SES): The Luxembourg based operator of *ASTRA*, Europe's largest *direct-to-home satellite (2)* system. (Website: www.ses-astra.com).

Society of Broadcast Engineers Inc. (SBE): This US Society was formed in 1963 as a non-profit or ganization serving the interests of broadcast engineers. It is the only US society devoted to the advancement of all levels of broadcast engineering. (W ebsite: www.sbe.org).

Society of Motion Pictur e and Television Engineers (SMPTE): US professional organization founded in 1916 with international branches that promotes interest in, generates and recommends standards for the television and film industries. (W ebsite: www.smpte.org).

soft: (1) An image which is not in sharp focus, optically or electronically . **(2)** In lighting, a source which gives a dif fuse light which does not cast a sharp shadow.

soft edge: A diffuse or graduated boundary to a picture image area. It is used in *keying* or *wipe* effects.

soft-edged masks: Graduated *neutral density* masks used in *slide mounts* to subtly blend adjacent projected images so that very lar ge *composite (1)* images can be made without visible joins.

soft focus: Producing an image with less than the maximum sharpness of which the system is capable.

softlight: A *luminaire* with an open bulb in a matt white *reflector* giving substantially shadowless illumination.

software: (1) The *program* or set of instructions which the *hardware* of a *computer* obeys to perform a task. **(2)** The audio and visual programme material used in *audio-visual* productions.

solarization: (1) Originally, a photographic ef fect in which the picture image is partially reversed in tone with light or dark edges at highlight and shadow boundaries. **(2)** A video effect where *luminance (2)* levels are inverted. The image therefore looks like a photographic negative.

solar outage: See *outage.*

solid state: An electronic device or system with no moving parts, heated filaments or gases, often applied generally to semiconductor technology.

solid state video r ecorder (SSVR): A digital video recorder using semiconductor storage in place of *magnetic tape* or *disk.*

soliton: A very narrow pulse of electromagnetic radiation which has a specially defined shape and frequency distribution, allowing it to travel very long distances in a transmission medium (e.g. *optical fibre*) with hardly any *dispersion (3).*

Sony Dynamic Digital Sound (SDDS): An 8-channel digital film sound system using *ATRAC* data compression at 5 :1. These tracks carry the data for 5 front channels (left, left centre, centre, right centre and right), two surround channels (left surround and right surround) and a *sub-woofer* channel. Masters are prepared using *DTRS (digital tape recording system)*. Two discrete tracks are recorded as pixels on both film edges (see **Appendix A**). The system was first used for the 1994 film 'Last Action Hero'. (Website: www.sdds.com).

Sony/Philips Digital InterFace (SPDIF): An interconnection standard developed by Sony and Philips for consumer digital audio equipment using *coaxial cable* and *phono* connectors. It is the domestic version of the *AES/EBU interface* and carries digital audio, with the addition of copy prohibiting and the inclusion of *compact disc subcode* data.

sort: To place in a specified order , usually *alphanumeric.*

sound drum: A roller used to control uniformity of movement of film where the sound track is read by a *scanning*-beam *slit (1)* or a magnetic *head (2).*

sound effects (SFX): Sounds added to dialogue in post-production, e.g. bird noises, engine noise, explosions, or a musical score.

sound gate: The *gate (1)* used instead of a sound drum to keep the film sound track aligned with the optical *scanning* beam.

sound head: (1) In motion picture projection equipment, a mechanism for moving the film and reproducing the photographic or magnetic sound record. **(2)** In audio or video equipment, the mechanism for recording/reproducing the audio information.

sound-on-film (SOF): A phrase used to indicate *synchronized* picture and sound on film. It was originally used to describe simultaneous recording of sound and pictures, e.g. in newsreels.

sound pressure level (SPL): Used as a suf fix to indicate a sound measurement.

sound reader: In film editing, equipment for monitoring an optical or magnetic *sound track.*

sound track: The defined area along the length of a recording medium such as film or *magnetic tape* which carries the audio information. Also, the sound record itself.

source code: A *computer program* written in a *high level language* which needs to be *compiled* or *assembled* before it can be used.

source intermediate format (SIF): A video picture format of 360 *pixels* by 288 lines at 25 Hz *frame rate* or 360 *pixels* by 240 lines at 30 Hz *frame rate.* It is the basic picture format used in *MPEG-1* systems.

space: One of the two states in a binary data transmission system, usually representing a binary nought (compare *mark*).

spacing loss: In magnetic recording, the loss of high frequencies due to imperfect contact between the replay *head (2)* and the magnetic medium.

spark: An electrician. (Colloq.)

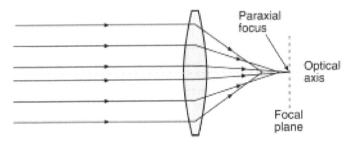

Figure S.7   Spherical aberration

spatial frequency: Effectively the fineness of a grid, measured in lines per millimetre.

spatial interpolation: See *interpolation (1)*.

spatial redundancy: Redundancy due to a lack of major changes across small areas within a *frame (2)* of a video sequence.

speaker support module: A *slide* sequence for use with a live presenter.

special effects (SFX): (1) A general term for scenes in which an illusion of the action required is created by the use of special equipment and processes rather than in reality , e.g. models, split screens.   **(2)** A video mixer enabling sections from two or more pictures to be montaged and displayed on the same *screen (1)*.

spectrum: (1) A range of *wavelengths* or frequencies of acoustic, visible or electromagnetic radiation. **(2)** Specifically a display of visible light radiation arranged in order of *wavelength.*

spectrum analyser: A piece of test equipment used to plot signal *amplitude* versus frequency. It displays the frequency distribution and levels of a signal across a band of the   *electromagnetic spectrum.*

specular: A reflection from a surface which is acting like a mirror . It will create a bright spot of light in the field of view of a camera.

speech track: A *voice track* as opposed to music or ef fects.

speed: A number indicating the sensitivity of a photographic film emulsion. See also *American Standards Association.*

spherical aberration: A lens *aberration* in which light rays passing through a lens near its edges come to focus at a point dif   ferent from those near the centre. It can be greatly reduced by   *stopping down* the lens. See **Figure S.7.**

spider: A three-armed *spreader* for *tripod* legs.

spigot: In theatre lighting, an adapter screwed to the hanging bolt of a *luminaire* case to enable it to be mounted on a floor -stand.

spike: A short unwanted *pulse* superimposed on an electrical signal, e.g. a switching *transient* carried through the mains *AC* supply.

spill: Light from luminaires which is not part of their main beams.

spill-over: A *satellite (2)* signal or other broadcast falling on parts of the Earth outside its defined coverage area.

spin: To rotate a video picture.

splice: A physical join in motion picture film or *magnetic tape.*

split edit: A video *editing* change of shot where sound and picture cuts occur at different times.

split focus: Setting focus on a camera to between two points of interest in the view to include both of them in the *depth of field.*

split screen: A shot in which two or more separate images appear in the same picture, separated by a *wipe pattern.*

splitter: A passive device (with no *active components*) which divides a signal into two or more paths.

spoilers: Pins used to prevent *head*-to-tape contact during rewind or fast-forward operations of a tape machine.

spoking: A distortion in a reel of film caused by loosely winding badly curled material.

spool: (1) A flanged hub on which film or *magnetic tape* is wound. **(2)** To wind a film or *magnetic tape* at a speed higher than the normal speed of reproduction.

spot beam: A focused *satellite (2)* transmission which only covers a small region of the Earth's surface.

spotting: (1) Locating individual sounds or words in a sound recording. **(2)** Retouching with an opaque medium to eliminate small transparent defects, such as pinholes, in the heavy *density* areas of a photographic image. **(3)** Marking *slides* to indicate their correct orientation for projection.

spot wobble: A small vertical oscillation of the *scanning* beam in large *screen (1)* television displays to render the *raster* spacing less obvious.

spreader: A triangular or Y-shaped device used on the floor to fix the legs of a *tripod*. Sometimes called a *crow's foot.*

spreadsheet: *Computer application* for working with numbers, able to calculate in columns and rows, and used mainly for financial statements.

sprite: In computer animation, a graphical figure which can be moved and manipulated as a single object. The term originally means a small mischievous pixie or elf.

sprocket: A toothed drum engaging with the *perforations* of film for *transport.*

sprocket holes: *Perforations* in motion picture photographic film. (Colloq.)

217

spyder

Figure S.8   Square wave

spyder: A low-level lighting support. (Colloq.)

square wave: A periodic electrical *waveform* switching virtually instantly between two voltage levels, with the 'on-off ratio' of equal or nearly equal duration. See **Figure S.8**.

squeeze: A digital video picture ef fect where the image is reduced in size.

squegg: An unwanted oscillation which leads to instability .

squelch: An electronic circuit in a *FM* radio *receiver* which *mutes* the output when the incoming radio signal falls below a pre-set *amplitude*. This prevents the listener hearing *noise* when tuning between stations.

stacker: See *projector stack.*

stage left: On the actor's left looking at the audience.

stage right: On the actor's right looking at the audience.

stage weight: A weight used to keep *french braces* steady when supporting scenery.

staggercasting: A *broadcast* programme is repeated at dif ferent times during the day.

stagger through: See *walk through*. (Colloq.)

staircase test signal: A video test signal *waveform* which rises or falls in equal steps between set limits. The modulated version is used to measure *differential gain* and *differential phase* errors. See **Figure S.9**.

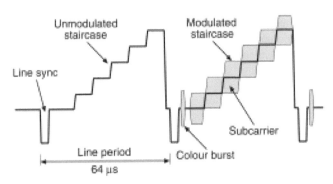

Figure S.9   Staircase test signal waveform

standard cell: A *primary cell* characterized by its constant *EMF* over a long period of time. It is used as a voltage reference.

standard definition: Refers to analogue and digital video having 525 lines or 625 lines per *frame (2)*.

standard play (SP): A recording mode allowing 2 hour recordings on E120 *VHS* tapes (compare *super long play*).

standards converter: Equipment to convert television pictures from one *scanning* standard to another, typically USA 525 lines to European 625 lines. The conversion will involve changing the number of lines and fields and the colour coding system.

standby: A waiting state, usually with reduced power or lamps of f.

star filter: A photographic *filter (1)* with an engraved or etched line pattern to produce star effects on bright light sources.

star network: A *network topology* where each *node (2)* is connected to a central *computer* which controls the system.

star trail: A digital video trail effect in which random *pixels* in the image trail are turned off, creating a blinking or starry appearance to the trail.

star wheel: The slotted *maltese cross* wheel in a *geneva* movement.

star wipe: A *wipe* in the form of a star . See **Figure W.1**.

start bit: In *asynchronous* serial data transmission, the first transmitted bit, preceding the first data bit. It informs the receiver that data is about to arrive.

state: One of the fixed angles in digital *phase modulation*.

statement: A *computer* term for a single *program* instruction in a *high level language*.

static: (**1**) Audible *noise* in radio or television receivers. (**2**) High voltages which can be generated by friction and may damage electronic components, particularly semiconductors. Abbreviated from *electrostatic*.

static marks: Marks caused by a discharge of static electricity on or near the surface of undeveloped photographic material which become visible after processing.

static memory: A semiconductor memory device where the information is retained without refreshing or recirculating so long as a power supply is applied to the chip.

static random access memory (SRAM): *Random access memory* where the information is retained without refreshing or recirculating so long as a power supply is maintained (compare *dynamic random access memory*).

station: A device on a *network* which controls the flow of information across the network.

stationkeeping: Carrying out minor orbital adjustments to a *satellite (2)* in *geostationary Earth orbit* to maintain its position within its allocated *box*.

statistical multiplexing: Also called 'stat mux', this technique allows a fixed data capacity channel to be used by several digital video or data services. A controlling *multiplexer (2)* allocates the data rate for each according to the complexity of the programme material.

status filters: *Filters (1)* of precisely defined transmission used in a colour *densitometer.*

Steadicam: An individual harness with a gyroscopically stabilized *camera mount.* It provides smooth action with a hand-held camera. (T rade name.)

step: To proceed through a *computer program* manually one instruction at a time.

stepped index fibre: *Optical fibre* in which the *core (2)* has a sharp but small change in *refractive index* at the join with the *cladding* (compare *graded index fibre*). See **Figure O.3.**

stepper motor: A type of motor which rotates in small but fixed steps, at any of which its position can be held.

step printing: Motion picture printing in which the film is exposed intermittently, frame by frame.

step wedge: (1) A camera test chart having rectangular areas of increasing lightness. **(2)** A test film image with a series of increasing density exposures used in processing.

steradian (sr): *SI unit* of solid angle, i.e. the three-dimensional angle at the tip of a cone as a proportion of a sphere.

stereo: Abbreviation for *stereophony* or *stereoscopy.*

stereophony: Sound reproduction using two or more channels to give the impression of width and depth which is absent from monophonic single channel reproduction.

stereoscopy: A system of photography giving the viewer a *three-dimensional* effect. It may use two separate images filmed from separate positions, one to be viewed by each eye.

stethoset: A type of headphone like a doctor 's stethoscope.

still frame: Continuous reproduction of one stationary image from a sequence of recorded images. The term is applied to both motion picture and video. See also *freeze frame.*

stock: The general term for unexposed cinematograph film. See also *raw stock.*

stock numbers: Sequential identification numbers applied at fixed intervals on the edge of cinematograph film during original manufacture.

stock shot: Also called a *library shot.* See *library.*

stop bit(s): In serial data transmission, one or more periods of equal duration to the data period which terminate a character or block. They ensure the receiver knows the current character or block has ended and can prepare for the next *start bit.*

stop frame: In *cinematography,* the repeated printing of a single *frame (1)* to appear as a still picture when projected.

stop motion: The operation of a motion picture camera, printer or *projector* one *frame (1)* at a time.

stopping down: Closing the *iris* of a lens to reduce the amount of light going through it. This reduces most *aberrations* and increases the *depth of focus.*

storage tube: A specialized *cathode ray tube* which does not need constant *refreshing* to maintain its display. (Hist.)

store-and-forward: The process by which video or audio sequences are digitized, data compressed and possibly edited to make a data file which can then be sent over a relatively low data rate circuit as a non- *real time* file transfer. The system is often used with a *satphone* for television or radio newsgathering in areas of the world where other techniques are not possible.

story board: A chart consisting of a series of still pictures, usually drawings, summarizing the contents of a proposed film, video or *slide* production.

strike: (1) To dismantle and remove *sets* or *props.* **(2)** To light an arc lamp.

stringy floppy: In *computer* systems, a bulk storage system using *magnetic tape.* (Hist.)

stripe: (1) A narrow band of magnetic material applied to photographic motion picture film for sound recording and reproduction. **(2)** The operation of *blacking* a videotape.

stripe mask: The shadowmask of a *slot mask tube.*

striping: (1) Applying a magnetic stripe to motion picture film. **(2)** Adding *longitudinal time code* to video *rushes (2)* after shooting. **(3)** See *blacking.*

strobe lighting: In film production, electronic flash lighting synchronized with a motion picture camera *frame rate,* giving sharp images of fast moving objects.

strobing: A disturbing effect in film or television arising from the field or *frame rate* distorting motion, particularly of rotating objects or stripes during a *pan.*

stroboscope: (1) A pulsed light source used to measure the speed of rotation or apparently freeze the motion of revolving objects. **(2)** A disc with equally spaced marks around its circumference used to check the rotational speed of objects, e.g. tape or turntable speed in the presence of *AC* lighting.

stylus: (1) A needle for *gramophone* disc reproduction, usually replace-able. May be made from sapphire or diamond. **(2)** In *computer graphics,* a pencil-like device for the input of information from a *data tablet* or for selection from a *menu*.

subcarrier: In the *PAL* and *NTSC* colour television systems, a sine wave with an extremely accurate frequency used to carry *colour-difference signals* within the normal video signal *bandwidth*. The *modulated* subcarrier is interleaved with the upper frequency components of the *luminance* signal. *Hue* information is carried as *phase modulation* and *saturation* as simultaneous *amplitude modulation* of the subcarrier.

subcarrier to horizontal (SCH): See *SCH phase*.

subcodes: Additional data capacity in a digital recording or replay format (e.g. *compact disc*) used to carry auxiliary information such as timing and track titles.

sub-quarter common intermediate format (SQIF): A video picture format with 128 *pixels* by 96 lines at 30 Hz *frame rate*, used for low *resolution* applications.

subroutine: A section of a *computer program (1)* to do a specified task. Subroutines may be called repeatedly from dif ferent parts of the main *program (1)*.

Subsidiary Communications Authorization (SCA): The name of the secondary channel used in stereo *broadcasts* in USA.

subtitle: A line or lines of words superimposed at the bottom of a film or video picture, either translating foreign dialogue or as an aid to the hard-of-hearing. See also *closed captioning*.

subtractive colour: Colour produced by removing unwanted colours from white light using transparent filters. The subtractive primary colours are *cyan*, *magenta* and yellow. This is the basis for colour film processing using dyes.

sub-woofer: A *loudspeaker* specially designed to reproduce very low audio frequencies, in the range 10 Hz to about 60 Hz.

sun gun: A portable hand-held lamp. (Colloq.)

super: See *superimpose*.

Super 16, Super 35: Film formats. See **Appendix A.**

super high frequency (SHF): The region of the *electromagnetic spectrum* between 3 GHz and 30 GHz. See **Appendix N.**

superimpose: (1) To add one picture on top of another, usually so that both continue to be visible. **(2)** To add a *caption* or *computer graphic* over a picture.

super long play (SLP): A recording mode allowing 6 hour recordings on E120 *VHS* tapes (compare *standard play*).

Super-Panavision: See *Panavision*.

superslide: A *slide* with a large image area of 40 × 40 mm, in a standard 50 mm square mount.

supertrouper: A very large follow spotlight.

Super-VHS (S-VHS): A *high-band* version of the *VHS* system using improved tape formulation. *Luminance (2) resolution* is improved, but colour *resolution* remains the same as normal *VHS*.

supplementary lens: An extra lens mounted on the front of a camera *objective lens*, e.g. a close-up lens.

supply reel: See *supply spool*.

supply spool: The reel on a machine handling photographic film or *magnetic tape* on which material is wound before passing through the rest of the path. Also called *supply reel* (compare *take-up spool*).

suppression: The reduction of generation of *radio frequency interference*, e.g. by fitting a resistor -capacitor circuit across switch or relay contacts.

surface: The upper or lower magnetically coated side of a *platter* in a *hard disk drive*. See also **Figure H.2**.

surface mount device (SMD): A small 'chip' component soldered directly onto *PCB lands*.

surround-sound: Sound reproduction with a multiple arrangement of *loudspeakers*, often four, providing sound apparently from all around the listener.

sweetening: (1) The compilation of complex sound ef fects to match the visual images. **(2)** In video, enhancement of a recorded image by electronic modification of tonal and colour rendering, edge sharpening, *gamma* correction, visual *noise* reduction, etc. **(3)** In audio, improving the subjective quality of an audio signal.

switched star: A form of *cable television* distribution.

switcher: (1) See *switched mode power supply*. (Colloq.). **(2)** US term for a *vision mixer (1)*.

symbol: In data communications, the basic unit of information transfer . The number of bits per symbol depends on the *modulation* system, e.g. 2 bit/symbol for *QPSK* and 6 bit/symbol for 64-QAM (see *QAM*).

sync: Synchronism or *synchronization*. See also *lip sync*.

synchronization: The fitting together in accurate time -relationship of video or film and its associated audio. Defined in standard SMPTE 318M (see **Appendix U**).

synchronizer: (1) In film editing, a group of two or more sprockets on a common shaft to allow lengths of perforated film to be wound through in a constant relationship. It may include a *frame counter* and a *sound reader*. **(2)** A digital device to store a *frame (2)* of video information which may be read out at a dif ferent rate to the input to provide a synchronous output.

synchronous digital hierar chy (SDH): A digital *telecommunications* standard for synchronous data transfers in a series of levels at data rates of up to 10 Gbit/s in 155 Mbit/s increments.

sync pulse: (1) In motion picture practice, a signal directly related to the speed of the camera, recorded on the magnetic audio tape for subsequent *synchronization*. **(2)** The part of a composite video signal that controls the repetition rate of the *scanning* system. Also called *line sync*. See **Figure C.5**.

sync-pulse generator (SPG): A stable source of *sync pulses (2)*, *colour black* and perhaps other timing pulses and video test signals. It may be a complete piece of equipment or be incorporated within other equipment such as video cameras, *videotape recorders*, *vision mixers*, etc. See also *genlock*.

sync tip: The most negative voltage level of the *sync pulses*.

Syndicat des Constructeurs d 'Appareils Radiorécepteurs et Télé-viseurs (SCART): The former name of the SIMAVELEC association of *audio-visual* manufacturers in France who defined the *SCART connector*.

synthesizer: (1) An electronic musical instrument used for producing sound effects or electronic music. **(2)** An electronic circuit that can accurately produce several different frequencies from a single reference frequency.

system clock reference (SCR): A data code sent at regular intervals in an *MPEG-1* system to *synchronize* the decoder's clock to the system clock.

Système International d'Unités (SI): The International System of Units. See *SI unit*.

T

tablet: See *data tablet*.

tabs: (1) In computing, tabulation points which allow text on a printer or display to be set in vertical columns. (2) Markers, often metal clips, attached to the edge of motion picture film to identify or cue a position. (3) Openable curtains on a stage. (The US term is *drapes*.)

tachometer: A *transducer* which converts speed of rotation into a scale reading or an electrical signal related to speed.

tail: The final section of a roll of film or tape.

take: (1) A scene or part of a scene photographed without a break in the action. See also *shot*. (2) See *cut (3)*. (3) A *vision mixer* control panel operation that puts a preset source to *air*.

take-up reel: See *take-up spool*.

take-up spool: The reel on a machine handling photographic film or *magnetic tape* on which material is wound after passing through the rest of the path. Also called *take-up reel* (compare *supply spool*).

talkback: A sound system allowing communication between the director, technical and production staff in studios.

tally light: See *cue light*.

tape: See *magnetic tape*.

tape guide: See *guide*.

tape path: The posts and *guides* that direct the *magnetic tape* inside audio and *videotape recorders*.

tape-slide: A sequence of *slides* synchronized with, or controlled by, an accompanying audio tape.

tape splice: A join in film or *magnetic tape* in which the ends are joined by adhesive tape.

tape streamer: A compact magnetic tape *cartridge* unit with a large data capacity often used for data back-up in a *computer* system.

tape transport: The mechanism for driving and guiding the *magnetic tape* in an audio or *videotape recorder*/reproducer.

target: (1) The surface of a video camera *tube (2)* on which the optical image is formed and scanned. (Hist.) (2) A small circular *flag (1)*.

T-bone: A low-level lighting support. (Colloq.)

teaching wall: An integration of facilities such as *flipcharts* and whiteboards with a front, rear or *overhead projector screen (1)*.

teaser: An overhead *gobo (2)* used to prevent *flare* in a camera arising from *back lights* or other light sources.

Figure T.1   Flying spot telecine

technical apparatus for the r ectification of indiffer ent film (TARIF): Equipment for adjusting colour reproduction when film is reproduced on a *telecine* system. See also *masking (6)*.

Technicolor: A motion picture colour film process developed by Herbert Thomas Kalmus, an American physicist. (T rade name.) (W ebsite: www.technicolor.com).

Technirama: A *widescreen* film production system using 35mm film photographed *anamorphically* at 1.5:1.

tele-: Greek for a long way , used as a prefix.

telecast: US term for a broadcast television programme.

telecaster: US term for a television broadcaster .

Telecine (TK): Equipment for replaying motion picture film as a video signal. See **Figure T.1** and also *flying spot scanner* and **Figure F.2**.

telecommunications: Communications over long distances by cables, *optical fibres, radio waves* or visual signalling.

Telecommunications Industries Association (TIA): A division of the *EIA* which represents manufacturers in the US *telecommunications* industry.

telecommunications satellite: A *satellite (2)* in space, usually in *geostationary Earth orbit* , which contains a set of *transponders* to provide *microwave radio-frequency* links between ground-based *trans-mitters* and *receivers*. The first commercial *telecommunications satellite (2)* was Early Bird in 1965.

teleconferencing: Electronically linking groups at different locations for a meeting. See also *video-conferencing*.

telegraphy: *Telecommunications* using interruptions or polarity changes in a DC current to transmit messages. First patented by Charles Wheatstone in 1837 then developed by Samuel Morse to use his code.

telephoto lens: A camera *objective lens* of long *focal length* with the secondary *principal point* in front of the lens (compare *retrofocus lens*). See **Figure P.10**.

teleport: A fixed *earth station* with several *dish aerials*.

Teleprompter: A television prompting device providing the script for an artist to read while looking directly at the camera. (T rade name.) See also *prompter*.

telerecording: Transferring a television or video presentation to motion picture film.

Teletel: French *viewdata* service.

teletext: A *videotex* data service, transmitting *alphanumeric* characters and simple graphics in the *vertical blanking interval* of *broadcast* video signals.

television (TV): (1) A system of transmitting electrical signals for the *real time* reproduction of a moving visible image at a distance, together with its associated sound. **(2)** A television receiver. (Colloq.)

television receive only (TVRO): Equipment for reception and display without transmission, often used for *off-air (1)* checks or *cues*.

Television South (TVS): The UK *Independent Television* broadcaster whose franchise covering the south of England was taken over by Meridian Broadcasting. (Hist.)

Telidon: A Canadian *teletext* service.

Telset: Finnish *viewdata* service.

temporal interpolation: See *interpolation (1)*.

temporal redundancy: Redundancy due to a lack of major changes between *frames (2)* in a video sequence.

Tentelometer: A tape tension measuring instrument. (T rade name.)

terabyte: 1024 *gigabytes*, equal to 1 099 511 627 776 bytes.

terminal: (1) A device remote from a *computer* at which data can enter or leave a *network*. **(2)** A small fixture for making an electrical connection.

termination: A *load (2)* inserted at the end of a *transmission line* to prevent the signal from reflecting back. Also called a *terminator*.

terminator: See *termination*.

tesla (T): *SI unit* of magnetic flux density, equal to one *weber* per square metre or 10 000 *gauss*. Named after the American physicist Nicola Tesla.

test film: A short film with standardized image sequences for testing optical alignment and resolution, picture steadiness and audio response of a film *projector* or *telecine* machine.

test tape: A pre-recorded *magnetic tape* for the *alignment* and testing of tape recorders/reproducers.

T-grain: A film emulsion containing tabular *grains*. A trademark of Eastman Kodak Co.

Thaumatrope: A disc with an image on each side, which appear to move when it is spun around a diameter . (Hist.)

thaw: The return to action after a *freeze frame* effect, especially in videowall presentation.

Theatre Equipment Association (TEA): Former name of the *International Theatre Equipment Association* (USA). (Hist.)

thermal magnetic duplication (TMD): A system of high-speed duplication of *videotapes* from a mirror-image master.

thermal printing: A printing system using a print head with several quick-heating elements and using heat -sensitive paper.

thermal trip: A protective device which mechanically interrupts the current in a circuit if it detects over -heating.

third harmonic distortion (THD): A measure of the unwanted contribution of the third *harmonic* in the *bandwidth* of a system (compare *total harmonic distortion*).

thermionic valve: An electronic device consisting of an evacuated glass envelope containing several metal electrodes and usually used as an *amplifier.* It was invented in 1906 by an American, Lee De Forest. See also *vacuum tube.*

thin film transistor (TFT): The switching structure used and hence an alternative name for an *active-matrix LCD* screen. Typically three transistors are used for each colour *pixel,* lying directly behind the liquid crystal cells they control.

thrashing: A computer term describing very high read and write usage of a *hard disk drive.*

thread: See *lace.*

three-perf: A system of 35mm motion picture photography using a *frame (1)* which is three *perforations* high instead of the standard four *perforations*. This reduces the length of film used for a given time by 25 per cent.

three phase: A three (or four) wire distribution of *AC* power supplies with the *waveform* in each of the three wires 120 degrees out of phase. The fourth wire in a star system is the neutral.

three-plane registration: In *slide projectors*, the precise positioning of a *slide* in three dimensions; vertical, horizontal and along the *optical axis*.

threshold howl: See *howl-round.*

through the lens (TTL): A camera *viewfinder (1, 2)* system where the image is obtained through the main *objective lens*, so that *parallax* error is eliminated. See also *reflex camera.*

throw: The distance from a *projector aperture (2)* to the centre of the *screen (1)*.

thumbwheel: A rotary control operated by moving the rim of an attached disc.

thyristor: A unidirectional controlled semiconductor switch often used as a power control device, e.g. in a lighting *dimmer*. Also called a *silicon controlled rectifier*. See also *triac*.

Tic-tac: *Viewdata* type system used in France.

tie line: An interlinking sound or vision connection cabled between technical areas.

tiling: Dividing a video image into rectangles of adjustable size, within which colour and *brightness* are integrated and displayed as average values.

tilt: (1) To rotate a camera in a vertical plane (compare *pan*). **(2)** To change the slope of the *frequency response* in a sound reproducer.

timebase: An electronic circuit generating timing signals. In a television display using a *cathode ray tube* it generates *deflection yoke* signals to make the spot scan across and down the screen to form the *raster*.

time-base corrector (TBC): An electronic device to correct timing errors in video signals, usually originated by fluctuations in speed in *videotape recorders*. A TBC is essential for replay into a studio or for transmission. It may be a stand alone unit for use with consumer videotape machines, or built into professional *videotape recorders*.

time code: A data signal recorded on audio and *videotape*, and sometimes on film, for subsequent *synchronization* and *editing*. It denotes a precise time in hours, minutes, seconds and *frames (2)* as a unique identification for each *frame (2)*. See also *longitudinal time code* and *vertical interval time code*. Defined in SMPTE 12M. See **Appendix U**.

time code in picture (TCIP): *Time code* numbers displayed over a video picture used in videotape *editing*.

time division multiplex (TDM): Sending multiple signals on one *channel (1)* by alternately assigning portions of each signal a period of time on the *channel (1)*. It was invented by AT&T in the USA in 1920.

time lapse: Recording a sequence of images with a controlled time delay between each picture, either on film or *videotape*.

timer: A mechanical, electromechanical or electronic device for controlling the timing of a sequence of events.

time shift: Recording a television *broadcast* programme for replay at a later time.

tint: (1) A colour mixed with white light, a desaturated colour, e.g. pink is a tint of red. **(2)** In the USA, sometimes used to refer to the *hue* of a colour.

tip: The end connection on a two- or three-circuit *jack* plug. See **Figure P.8**.

tip projection: The penetration depth of a video *head (2)* into the *videotape*. It is measured as the protrusion of the video *head (2)* outside the drum.

titles: Words appearing in a motion picture or television production which do not form part of a scene.

T-number: A calibration system for the light transmission of a motion picture camera lens. It is based on its actual transmission at various diaphragm settings. This means that two lenses with the same T-number will always give the same image brightness (compare *f-number*). Also called *T-stop*.

Todd-AO: A *widescreen* film production and projection system using a 65mm *camera original* and 70mm *release prints* with a stereo magnetic sound track. (Hist.)

toggle: To clock from one stable state to the other .

token ring: A *ring network* where a special character sequence (token) is passed between the *nodes*, which can only send data when they have received the token.

tone: (1) The variation in a colour, or in the range of greys between black and white. (2) A constant frequency audio signal.

tone arm: See *pick-up arm*. (Hist.)

tone control: An electronic circuit for manually altering the *frequency response* of *amplifiers*, commonly giving boost or cut in treble and/or bass.

tongue: US term for *jibbing*.

top hat: A small *camera mount* used when a very low position is required. Also called a *hi-hat*.

top light: A *luminaire* directly overhead.

topology: The physical layout of a *network*.

toroidal transformer: A high efficiency *transformer* characterized by its donut shape and its consequent immunity to external magnetic fields and minimal generation of interfering fields.

TOSLINK: An optical *interface (1)* carrying *SPDIF* formatted data, available on some domestic audio equipment.

total harmonic distortion (THD): A measure of the *distortion* of a signal. It is the power ratio of the *harmonics* of a pure fundamental frequency to the fundamental, usually measured in *decibels*.

touch button: An electrical switch which operates by capacitance or other effect, i.e. with no physical contact closure.

touch screen: A visual display from which an item may be selected by physically touching that part of the *screen (1)*.

track: (1) A defined part of the width of a photographic or magnetic recording medium which carries discrete information. (2) A pair of parallel rails on which a camera *dolly* runs. (3) In a *hard disk drive* or *floppy disk*, one complete circle on a ﾠﾠsurface on which data can be recorded. See also **Figure H.2**. (4) A *circuit* interconnection etched or printed onto a *printed circuit board*.

tracker ball: A two-dimensional control in the form of a rotating ball in a socket, used in *computer graphics* to control the position of a *cursor*.

tracking: (1) Physical movement of a camera and its mount towards or away from the subject, or to follow a moving action. It produces a different visual ef fect to a *zoom*. (2) In electronics, a narrow band electronic *filter (3)* which automatically alters its centre frequency according to the frequency of the incoming tone. It may be used to follow changes of the fundamental tone or its harmonics. (3) In *audio disc* replay, the path of the pick-up *head (1)* following the grooves or other method of location. (4) In *videotape recorders*, causing the video *heads (2)* to follow exactly the recorded *video tracks*.

traffic: The data flowing across a *network*.

trailer: (1) A short film advertising a forthcoming cinema presentation or a similar technique in television. (2) The identification and protective leader at the end of a reel of motion picture film. (3) The optional last component in a data *packet*.

trailing edge: The second falling or rising edge of a pulse (compare *leading edge*).

tranny: Abbreviation for *transparency, transformer* or transistor. (Colloq.)

transcoder: A system to convert video from one colour standard to another, e.g. *PAL* to *SECAM*.

transcription: (1) A written text from a sound track. (2) A high-quality *audio disc* turntable. (Hist.)

transducer: A device for converting between electrical, magnetic, acoustic or mechanical energy.

transformer: A device for changing the voltage of an *alternating current* signal.

transient: A momentary disturbance of an electrical signal.

transition: A visual change from one *shot* to another using a *cut (2)*, *mix (2)* or *wipe*.

transmission (TX): (1) The information sent by a *transmitter*. **(2)** The amount of light emer ging from a projected photographic *slide* or film. See also *density*.

transmission control protocol (TCP): A network transmission *protocol* providing reliable, connection-oriented, *full-duplex* data streams. It uses IP (*Internet protocol*) for delivery.

transmission control protocol/Internet protocol (TCP/IP): A commu- nications *protocol* system originally developed by the US Department of Defense which has become the de facto standard for transmitting data over networks, in particular the *Internet*.

transmission line: The physical media used to transmit signals between two devices.

transmitter: A device which sends information either in the form of an electrical signal (e.g. a *transducer*) or an *electromagnetic wave* (e.g. in radio or television systems).

transparency: An image to be viewed by transmitted light, either directly or by projection. See also *slide.*

transparency viewer: A device containing a diffuser and a lens (sometimes with a self-contained light source) for viewing a *transparency.*

transparent: (1) Allowing the passage of light or other *electromagnetic* energy with minimal distortion and attenuation. **(2)** In video, a condition where the output of a system is identical to the input. **(3)** In *computers*, a condition where the action of part of a system is effectively undetectable by the person or system using it.

transponder: (1) A transmitting/receiving system which will respond with its own signal when interrogated. **(2)** A receive and transmit *channel (1)* on a *satellite (2)* incorporating a high-gain amplifier , a frequency changer and a *high-power amplifier.*

transport: The physical mechanism controlling the movement of tape or film in an audio or *videotape recorder*/reproducer or film camera or *projector.* Also used to describe the movement of film or tape through the mechanism.

Transportable Earth Station (TES): A mobile *earth station*, typically used for *satellite news gathering* operations.

transport stream (TS): A *MPEG-2* digital *multiplex (2)* of video, audio and data associated with one or more services and optimized for transmission of material in an error-prone environment, e.g. *satellite (2)* broadcasting. It uses fixed length 188 byte *packets* to which are added 16 bytes of *error correction codes.* The final transmitted data stream is optimized for cable, terrestrial or *satellite (2)* distribution (compare *program stream*).

transposer: A television *repeater* that does not fully demodulate the received signal before transmitting it.

transverse scan: An early method of analogue *videotape* recording. The tape is curved across its width and tracks are recorded across the tape by magnetic *heads (2)* on a spinning *head wheel* at 90 degrees to the tape motion (compare *helical scan*). (Hist.)

transverse waves: Waves in which the displacement is at 90 degrees to the direction of propagation, e.g. water , light and radio waves (compare *longitudinal waves*).

trapezium distortion: Image distortion in an optical or video system causing a rectangle to appear with one or more sides converging unequally on each another (compare *keystone distortion*).

trapezoidal distortion: See *keystone distortion.*

travelling matte: In *cinematography,* a special effects shot in which the foreground action, usually photographed against a blue backing, is superimposed on a separately recorded *background (3)* by laboratory printing processes using *mattes.*

travelling peg bars: *Peg bars* recessed into an *animation table* which can be moved mechanically in an east/west direction.

travelling wave tube (TWT): A powerful *microwave* amplifying *valve* used in a *high-power amplifier* in *satellite news gathering* systems and on communications *satellites (2).*

tray: US term for a *slide magazine (2).*

tree and branch: A form of cable distribution system.

tremendously high frequency (THF): The region of the *electromagnetic spectrum* between 300 GHz and 3000 GHz. See **Appendix N.**

triac: A bidirectional controlled semiconductor switch often used as a power control device, e.g. in a lighting *dimmer* (compare *thyristor*).

triangle: A three-sided *spreader* for *tripod* legs.

triax: A *coaxial cable* with an extra *screen (3)* capable of transmitting power and coded commands to a television camera as well as receiving the video signals from the camera. It is used in television studio installations and on *outside broadcasts,* particularly when the camera is a long distance from its *camera control unit.* See also *camera cable.*

trigger: (1) To initiate an action. (2) Mechanical lever to start or stop an action. (3) Electronic pulse of short duration to operate a circuit. (4) The electronic initiation of an *oscilloscope* display.

trims: The portions of film for a scene which are left over after the selected section has been used in the final assembly of a motion picture film.

Triniscope: A video recording system using separate *cathode ray tubes* with reflectors and *dichroic* mirrors to combine the red green and blue images for recording on film. (Hist.)

Trinitron: A colour television *cathode ray tube* using phosphor stripes and a striped *aperture grille.* (Sony trade name.)

tripack: General term for photographic material having three layers of sensitive emulsions for recording and reproducing colour images.

triple standard: A video picture monitor , *videotape recorder* or other equipment, capable of accepting video signals of *NTSC, PAL* or *SECAM* standard.

tripod: A three-legged equipment support for a camera, portable *screen (1),* etc., which is adjustable in height.

trombone: A lighting support for a lamp on a scenery *flat* within a set. (Colloq.)

trouper: A follow spotlight. (Colloq.)

truck: (**1**) See *tracking (1)*. (**2**) See *OB truck*.

trucking: See *crab*.

T-stop: See *T-number*.

tube: (**1**) US term for a domestic television *receiver*. (Colloq.) (**2**) An image sensor which used to be used in video cameras and *telecine* machines. See *Plumbicon*. (Hist.)

tumbling: In *computer graphics*, moving a three-dimensional object on a display by continually changing its axis of rotation.

tungsten-halogen lamp: An efficient lamp consisting of a tungsten filament in a quartz or hard-glass envelope containing a halogen (bromine or iodine). This avoids internal blackening of the envelope by the deposition of tungsten from the filament. See also *quartz iodine lamp*.

tungsten lighting: Lighting with incandescent lamps having tungsten filaments, including *tungsten-halogen lamps*.

turnkey: A complete system or installation which has been supplied self-contained and ready to use, e.g. a *computer* with both the *hardware* and the *software* required for a particular application.

turntable: (**1**) A motorized constant-speed rotating support for a disc, typically a *gramophone* record. (**2**) A mechanism fitted to an *animation table* or camera to produce *rotations*. (**3**) A rotating circular platform in a studio.

turret: A rotatable mount on a film or video camera allowing any one of two or more lenses to be brought into position on the *optical axis* as required. Superseded by *zoom lenses*. (Hist.)

turtle: A low-level lighting support. (Colloq.)

tweek: A small engineering adjustment. (Colloq.)

tweeker: A small screwdriver, sometimes with a non-conducting blade, used to make adjustments to electronic equipment. (Colloq.)

tweeking: Making small technical adjustments to a device or system. (Colloq.)

tweeter: A *loudspeaker* designed to reproduce only high audio frequencies.

twinkle: In *slide* projection, an effect giving rapid fluctuations in image intensity.

twinning stand: A stand which supports one automatic *slide projector* on top of another.

twisted pair: A signal connection consisting of two insulated wires twisted together and used for telephone connections and many *computer* network connections. The twists reduce *crosstalk* with other pairs in the same cable or duct. See also *shielded twisted pair*.

twitter: A video picture impairment where horizontal borders of objects in the picture flicker due to *interlace*.

two-duration control: The control of two *slide projectors* to give fast and slow *dissolve* changes by means of *cue tones* of 1000 Hz but of two different durations, typically 50 ms and 450 ms.

U

U: (1) *Colour difference signal* used in PAL. (B–Y) reduced in *amplitude* and *bandwidth* limited to about 1.2 MHz. U = 0.493 (B –Y). **(2)** Rack-height unit. A measure of the height of *equipment racks*, where 1 U = 44.5 mm (1.75 inch).

U-format: The *video cassette* format using 19 mm (¾ inch) tape compatible with the *U-matic* system.

U-link: U-shaped *musa (1)* device for vertical connections on a video *jackfield*. Called a *hairpin* in the USA. See **Figure U.1**.

Figure U.1 U-link

ultra high frequency (UHF): The region of the *electromagnetic spectrum* between 300 MHz and 3000 MHz. See **Appendix N**.

ultra low frequency (ULF): The region of the *electromagnetic spectrum* between 300 Hz and 3000 Hz. See **Appendix N**.

Ultra-Panavision: See *Panavision*.

ultrasonic: Sound at frequencies above the upper audible limit for humans.

ultrasonic cleaner: A machine for cleaning motion picture film by passing it through a solvent agitated at *ultrasonic* frequencies.

ultraviolet light (UV): Invisible electromagnetic radiation at *wavelengths* shorter than the violet end of the *visible spectrum* (about 400 nm) and longer than *X-rays*. It is detectable by photographic means, fluorescence, or by suitable photo-diodes or photo-transistors.

U-matic: A *video cassette* system using 19 mm (¾ inch) tape for corporate, industrial and *ENG* applications, now obsolete. (Sony trade name.)

unblooped: A *splice* in a photographic sound track which has not been *blooped* and which may therefore produce a *noise* on reproduction.

unbalanced: Referring to a two-wire circuit with one connection at or near *ground* potential. Any *interference* picked up along the cable will distort the signal being sent. See also *ground loop* (compare *balanced connection*).

uncommitted bit rate (UBR): An *ATM* traffic pattern where the data service is only allocated spare capacity on the network which no-one else needs. See also *available bit rate (ABR), current bit rate (CBR)* and *variable bit rate (VBR)*.

under-cranking: Running a film camera at less than its normal speed to speed up an action on projection.

under-damped: *Damping* in which a steady state is reached only after several oscillatory cycles. See **Figure D.1.**

underscan: Reducing the displayed television picture size by about 10 per cent to allow the picture edges to be seen.

unidirectional: Sensitive predominantly in one direction.

unidirectional microphone: See *directional microphone.*

uniform (or universal or unique) r esource locator (URL): A string of characters (including a *website* address) that uniquely identifies resources on the *World Wide Web.*

unique resource locator (URL): See *uniform resource locator.*

United News & Media plc (UNM): An international media and information group based in the UK. (W ebsite: www.unm.co.uk).

unit interval (UI): In *telecommunications*, the duration of one bit period at the specified *bit rate.*

Universal Mobile Telecommunications System (UMTS): The third generation (3G) of mobile telephone technology , which allows mobile phones to carry data and *Internet* services, including the display of moving images.

universal resource locator (URL): See *uniform resource locator.*

universal serial bus (USB): A *peripheral* interconnection system for *PCs.*

Universal Time Co-ordinated (UTC): The standard for measuring time, locked to an atomic clock reference. The same as *Greenwich Mean Time.* Also called *Co-ordinated Universal Time (CUT).*

Unix: A general purpose multi-user , multi-tasking *computer operating system* developed in the 1970s which can be used on dif ferent computer *platforms.*

unmod: Unmodulated. A photographic sound track on which no signal has been recorded.

unsqueezed: A motion picture print in which the compressed image of an *anamorphic* negative has been corrected for normal projection.

upconverter: Electronic equipment used to increase the number of television lines from *standard definition television* to *high definition television* (compare *downconverter*).

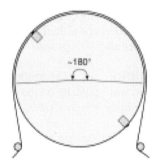

Figure U.2   U-wrap

uplink: In a *telecommunications satellite (2)* system, the outward path to the *satellite (2)* from the transmitting *earth station's dish aerial* (compare *downlink*). Sometimes used to describe the transmitting *earth station* itself.

up rate: The speed of the up-coming lamp in a *slide projector dissolve* change system.

up stage: Performing area furthest from the audience or camera.

user datagram pr otocol (UDP): An *Internet* system transport *protocol* using IP *(Internet protocol)* for delivery but, unlike TCP *(transmission control protocol)*, *datagrams* can be exchanged without acknowledgements or guaranteed delivery.

utility: A small computer program which performs maintenance or general purpose chores in the system.

U-wrap: The *wrap* around the *head drum* of a *helical scan magnetic tape* recorder giving just more than 180 degrees of contact, therefore requiring at least two rotating *heads (2)*. See **Figure U.2** (compare *omega wrap*).

V

V: *Colour difference signal* used in PAL. (R–Y) reduced in *amplitude* and bandwidth limited to about 1.2 MHz. V = 0.877 (R –Y).

vacuum tube: An evacuated glass envelope containing several metal electrodes. It was invented first as a diode by Sir JohnAmbrose Fleming. See also *thermionic valve.*

valve: See *thermionic valve.*

valve rollers: See *fire rollers.*

variable area track: A photographic sound track in which the image width varies according to the sound *modulation.*

variable bit rate (VBR): (1) Networked data that is tolerant of delays and changes in its allocated data rate. It is an *ATM* service class that uses *statistical multiplexing.* See also *available bit rate (ABR)* , *current bit rate (CBR)* and *uncommitted bit rate (UBR)* . **(2)** In video and audio data compression, optimizing picture or sound quality by varying the bit-rate according to image or sound complexity . This either saves storage space, e.g. on *DVDs,* or allows more efficient allocation of the total bit-rate in a broadcast *multiplex (2)* (compare *constant bit rate (CBR)*).

variable density track: A photographic sound track in which the *density* of the image varies according to the sound *modulation.*

variable-frequency control: A control system for a pair of *slide projectors* using a continuous tone of varying frequency . Also called *FM control.*

variable length coding (VLC): A digital coding technique in which more frequently occurring input sequences are represented by shorter codes, and less common sequences are replaced by longer codes. This form of coding is often used in video data compression schemes such as *JPEG, MPEG* and *DV.* Morse code is an example of VLC. The letter E is given as a single 'dit' whereas Q is given the code 'dah, dah, dit, dah'. See also *Huffman coding.*

variable transformer: A *transformer* with a smoothly adjustable output voltage.

Variac: A *transformer* with a continuously variable output voltage. (Trade name.)

variator: The lens or group in a *telephoto lens* which moves to change the image size (compare *compensator*).

varifocal lens: An alternative term for a *zoom lens.* (Hist.)

varispeed: A continuously variable speed control on analogue magnetic audio recorders. It usually provides a wide range, but may be a *pitch control.*

V-chip rating: A USA television programme rating scheme required to be fitted in all television receivers with 33 cm (13 inch) or lar ger screens. The 'V-chip' reads information encoded in the rated programme and blocks programmes from being displayed on the set based on the rating selected by the parent. (W ebsite: www.fcc.gov/vchip).

vector: (1) A representation of magnitude and direction. (2) A line connecting defined points on a *cathode ray tube*. (3) In computers, a pointer used to allow access to a relocatable *program* or data.

vector graphics: A method of presenting graphical data with the display composed of *vectors (2)* drawn directly from point to point on the *cathode ray tube* instead of using a *raster*, thus avoiding *aliasing*. Also called *vector mode display*.

vector mode display: See *vector graphics*.

vectorscope: Television measuring and signal monitoring equipment for displaying colour information. It allows the colour attributes being displayed to be monitored and assessed.

venetian blinding: See *Hanover bars*.

vertical blanking interval (VBI): See *vertical interval*.

vertical interval: The period in a video signal between the end of one displayed scanning *field* and the start of the next during which the scanning spot in a *CRT*-based picture display device returns from bottom to top. Also known as the *VBI*. See also *blanking period* and **Appendix L**.

vertical insertion test signal (VITS): See *insertion test signal*.

vertical interval time code (VITC): A *time code* signal carried on two non-adjacent lines in the *vertical interval* of a video signal between each field of the *active picture*. Its advantage over *LTC* is that it can be read at *still frames* and very low tape speeds.

vertical polarization: See *linear polarization*.

very high frequency (VHF): The region of the *electromagnetic spectrum* between 30 MHz and 300 MHz. See **Appendix N**.

very low frequency (VLF): The region of the *electromagnetic spectrum* between 3 kHz and 30 kHz. See **Appendix N**.

vestigial sideband (VSB): Type of modulation used, for example, in *PAL* transmission and the American ATSC *(Advanced Television Systems Committee)* digital television standard.

VHS-C: A compact *video cassette* using the *VHS* format. An adapter is needed to allow replay in standard *VHS* equipment.

VIBGYOR: Violet, indigo, blue, green, yellow , orange, and red. A mnemonic for the order of colours in the *visible spectrum*, i.e. outside this range are *ultraviolet* and *infra-red*. See also *ROYGBIV*.

video: (1) Electronic means of recording, generating, storing and reproducing visual images. (2) A home *video cassette recorder*. (Colloq.) (3) A *videogram*. (Colloq.) (4) See *video signal*.

video cassette: A plastic cassette containing *videotape* with separate supply and take-up *spools*. In the recorder or player , a loop of tape is drawn from the cassette into the machine (compare *cartridge* or *reel-to-reel*).

video cassette recorder (VCR): The name used for a video recorder that accepts *video cassettes*. It usually refers to a consumer or industrial quality machine, with *VTR* being the preferred term for *broadcast standard videotape recorders*.

Video CD: A *CD* containing digital video and audio compressed to the *MPEG-1* standard. It is defined in the *White Book*.

video compact cassette (VCC): Obsolete consumer quality *video cassette recorder* system developed by Philips using 12.5 mm ($\frac{1}{2}$ inch) tape. (Hist.)

video conferencing: The use of bi-directional vision and sound links to permit a conference between two or more participants at different locations. The picture is usually updated at less frequent intervals than in broadcast video to reduce the transmission *bandwidth* required. See also *teleconferencing*.

videodisc: A video and audio recording on a flat circular medium, usually using digital storage, by optical (*laserdisc*) or mechanical (capacitive disc, now obsolete) means.

video distribution amplifier (VDA): A video signal *amplifier* designed to give a substantially flat *frequency response* over the video band and provide several separate output signals. It may also perform cable equalization.

video dub: Recording or re-recording the *video track* of a *videotape* recording without disturbing the existing audio signal.

Video Electronics Standards Association (VESA): An association which promotes and develops visual display and display *interface (1, 2)* standards. (Website: www.vesa.org).

videogram: The programme content of a *video cassette* or *videodisc*. Often simply called a *video*.

videographer: A video camera operator. (Colloq.)

videographics: Using a computer to manipulate video images.

videography: The art and practice of making video films.

video high density (VHD): A grooveless capacitance *videodisc* system. (Trade name.) (Hist.)

video home system (VHS): A consumer quality *video cassette recorder* system developed by the Victor Company of Japan which uses 12.5mm ($\frac{1}{2}$ inch) tape. (Trade name.)

videophone: (1) Still picture transmission over conventional telephone lines. (Hist.) **(2)** Transmission of low resolution video images over conventional telephone circuits. Also called *video telephone*.

video server: A *server* system using an array of high capacity *hard disk drives*, often interconnected with other systems using *FibreChannel* and used to store video programmes, commercials or promotions. The more general term is *media server.*

video signal: The electrical signal representing the images in a television system.

videotape (VT): (1) *Magnetic tape* specifically designed for use as a video recording medium. **(2)** The operational area where video recordings are made and replays originated.

videotape recorder (VTR): A machine to record and replay television pictures and sound on *magnetic tape.* Usually used when referring to *broadcast standard* video recorders.

video telephone: See *videophone.*

videotex: A general term for an information service of selected *alphanumeric*/graphic data, presented as a still-picture video display, e.g. *teletext.*

Videotext: A German *teletext* system.

video track: The area on a *videotape* where picture information is recorded.

videowall: A mosaic assembly, usually in a rectangular block, of a number of individual video *screens (1)*, displaying multiple, individual or composite images from *VTR, videodisc* or *computer* sources.

Viditel: Dutch *viewdata* service.

viewdata: A *videotex* data service distributed by the public telephone network and accessed by direct dialling.

viewer: (1) An optical device for inspecting the picture image on a film or *slide.* **(2)** In *desk-top editing*, a *window (2)* which allows video clips to be played and viewed. **(3)** One who watches television broadcasts(!)

viewfinder: (1) In video production, a small picture monitor allowing the video camera operator to frame and focus the scene. **(2)** In film production, an optical system allowing a film camera operator to frame and focus the scene. It may be *through the lens* and give the operator the same view as the camera lens. If it is separate, a *parallax* error will be introduced.

vignetting: (1) A gradual shading towards the edges of an image. In a lens system, it is usually caused by mar ginal light near the edges of a lens system being partly obscured by the lens barrel. Also called *portholing (1).* **(2)** Producing an image with a *sharp* centre and deliberately blurred edges.

vinegar syndrome: Decomposition of *acetate*-base film, producing a characteristic odour.

virtual reality: A video technique using sophisticated *computer graphics* to generate three-dimensional images which may be manipulated in real time to create an artificial environment.

virtual studio: An artificial environment created in a *computer* system into which actors and objects in a real environment can be added using *chromakey.* Movements of the video camera imaging the actors and objects are sensed and converted into corresponding movements in the virtual studio, giving a 'realistic' impression of it being a genuine physical scene.

virus: See *computer virus.*

viscous process: Photographic processing in which the chemicals are applied as layers of viscous solutions.

visible spectrum: The range of colours which can be seen by human observers, corresponding to electromagnetic radiation with wavelengths between about 400 nanometres (for violet) and 700 nanometres (for red).

vision apparatus room (VAR): An equipment room near to a television studio housing the technical equipment associated with the studio, e.g. camera *CCUs, vision mixer (1)*, etc.

vision mixer: (1) Equipment which allows the selection and combination of video sources in a television studio, editing or transmission operation. Called a *switcher* in the USA. **(2)** The operator of a vision mixing console.

Vistamorph: A 70 mm *widescreen* format originated in 35mm *anamorphic VistaVision* and *Technirama* and printed to 70mm five *perforation* vertical format for *anamorphic* projection to give a 3 :1 *aspect ratio.* It can also be printed to 35mm four *perforation* vertical format for *anamorphic* projection at 3 :1 *aspect ratio.*

VistaVision: A film origination format with the film being transported sideways through the camera, eight pairs of *perforations* at a time. The proportions of the full frame are 1.5 :1 but it is normally regarded as a 2:1 *aspect ratio* format with 1.85:1 *release prints.*

visual display terminal (VDT): Used for communicating with *computers.* (Hist.)

visual display unit (VDU): A display device containing a *cathode ray tube,* on which information, often from a *computer,* can be shown.

Visual Information Processing System (VIPS): An *interactive computer/videodisc* system. (Sony trade name.)

voice coil: The cylindrical coil inside a slot in the permanent magnet of a *loudspeaker.* When a current representing an audio signal is fed through the coil it moves the attached cone back and forth to create sound energy.

voice over: A narrative recording on a programme which is heard without the speaker being seen.

voice track: A recording track reserved for narration or dialogue as opposed to music or ef fects. Also called a *speech track.*

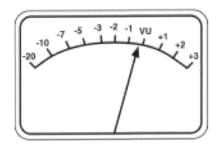

Figure V.1   VU meter scale

volatile memory: Data memory where the data is lost if power fails or is switched off.

volt (V): *SI unit* of *electromotive force,* equal to the difference in electrical potential able to send a current of one *ampere* through a conductor with a resistance of one *ohm.* Named after the eighteenth century Italian scientist Conte Alessandro Volta.

voltage controlled amplifier (VCA): An *amplifier* in which the gain is proportional to a control voltage. It can serve as an *automatic gain control.*

voltage controlled attenuator (VCA): A less common meaning for the acronym VCA. An attenuator in which the attenuation is proportional to a control voltage.

voltage controlled oscillator (VCO): An oscillator in which the frequency is proportional to a control voltage. It can serve as an *automatic frequency control.*

volt-ampere (VA): A measurement of AC power. The product of voltage and current in an AC circuit, without regard to phase angle. Due to inductance and/or capacitance altering the voltage/current phase angle, VA may not be the same as watts.

volt-ohm meter (VOM): An alternative term for a *multimeter.* (Hist.)

volts, alternating current (VAC): An electric voltage which reverses its polarity at regular intervals, e.g. the mains supply of 220 VAC at 50 Hz in Europe, 110 VAC at 60 Hz in USA.

volts, direct current (VDC): An electric voltage which has a fixed polarity, e.g. a 9 VDC battery.

volume: The intensity or magnitude of sound.

volume unit (VU): A unit of sound level measurement defined as the percentage utilization of the *channel (1).*

vox pops: In television programmes, short interviews with randomly chosen members of the public.

VPB: An open Quantel standard file format in which images, browse, stencils and cut-outs are transferred. It is based on the *ITU-R BT.601* specifications.

VU meter: Volume unit meter. An audio level meter with defined ballistics and an average rectifier characteristic. It indicates the mean enegy of the signal. See **Figure V.1**.

W

walk through: A full rehearsal including camera movements, but without actually filming or recording sound. Also called a *dry run* or *stagger through*.

wall sled: A lighting support for a lamp on a scenery *flat* within a set. (Colloq.)

wand: A hand-held *barcode* reader.

warble tone: A constant amplitude tone used in audio testing whose frequency varies between set limits, i.e. it is frequency *modulated*.

warm start: To *reboot* a system without turning of f the power (compare *cold start*).

wash: A general flood of light covering the setting.

watt (W): *SI unit* of electrical power, equivalent to producing or expending one *joule* of energy per second or the power used when one *ampere* flows through a one *ohm* resistor. Named after the eighteenth century Scottish engineer and inventor James Watt.

waveform: A presentation of the varying *amplitude* of a signal in relation to time.

waveform monitor: A specialized *oscilloscope* for displaying and measuring the *waveform* of video signals.

waveguide: A hollow metallic conductor , usually rectangular in shape, used to carry *microwave* signals between transmission/reception equipment and dish *aerials*. The *electromagnetic waves* are confined within the dimensions of the conductor .

wavelength: The distance between successive corresponding points in a wave.

wavelength division multiplexing (WDM): Simultaneously transmitting different wavelengths in an *optical fibre*.

weave: Periodic horizontal unsteadiness of a projected film image (compare *hop*).

weber (Wb): *SI unit* of magnetic flux, equal to 100 million *maxwells*. Named after the nineteenth century German physicist Wilhelm Eduard Weber.

website: An addressable *computer* on the *Internet* which contains a collection of *WWW* information in the form of *HTML* pages.

weighting: (1) A correction factor applied to a signal or measurement, e.g. a reduction in amplitude of the raw *colour difference signals* to produce *Cb* and *Cr*. **(2)** In audio, a standardized modification of the *frequency response* of measuring instruments for *noise*, *rumble* and *wow* so that the indications correspond to subjective ef fects.

wet printing: See *liquid gate.*

what you see is what you get (WYSIWYG): The final thing looks exactly the same as it does on a *monitor (1).*

whip pan: A fast *pan* which blurs the image. In *animation,* whip pans have the artwork reduced to blurred lines between the initial and final images.

White Book: The standard for *Video CD.*

white clipper: A video circuit which clips of f any signal rising above a specified video level.

white crushed: A picture condition where an increase in *white level* results in loss of gradation in light areas.

white level: The *amplitude* of a video signal representing the maximum allowed *exposure,* i.e. the lightest part of the picture. See also *peak white.*

white noise: *Noise* containing uniform energy distribution over its entire frequency range.

wide-angle lens: An *objective lens* of short *focal length.* See also *retrofocus lens.*

wide area network (W AN): The interconnection of *computers* and *peripheral* equipment over a lar ge geographical area, e.g. a whole country.

widescreen: In general, pictures presented with an *aspect ratio (1)* greater than 1.4:1. In television this is usually 16 :9. Not to be confused with *letterbox presentation.*

wide screen signalling (WSS): A digital signal during the first half of line 23, used in some analogue 625-line television systems to indicate the *aspect ratio (1)* and other information about the transmission. It is used in analogue television broadcasts in the *PAL plus* standard in some European countries.

wild shooting: Pictures recorded without synchronized sound.

wild track: Sound recorded without a simultaneous picture, used to capture the ambient background sounds during the filming.

Winchester disk: A sealed *hard disk drive* developed by IBM in the 1960s for storing large amounts of data with a reasonably fast access time, i.e. faster than a *floppy disk* but slower than *solid-state* memory. Originally the 30–30 drive (30 Mbyte fixed and 30 Mbyte removable storage), it was nicknamed after a Winchester rifle carrying the same 30 –30 designation. (Hist.)

windshield: An attachment to reduce wind noise and *blasting* in a *microphone.*

window: (1) US term for a *pulse and bar* test signal giving a lar ge white square in the centre of the picture. **(2)** In *computers,* a specific area on the screen through which part of an image or data file can be viewed.

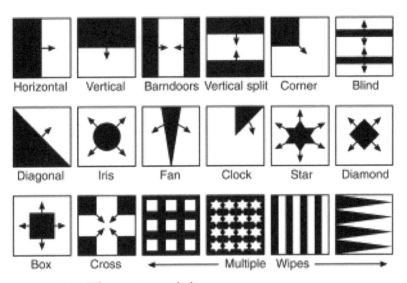

Figure W.1   Wipe pattern symbols

window dub: US term for a copy of a *videotape* with *time code* numbers keyed into the picture.

windows: *Computer* technique for presenting information from more than one source visually on a display screen. Some of the data appears in a *window (2)*, which may be varied in size.

windows, icons, mouse and pull-down menus (WIMP):   In computing, the most common type of *interface (2)* between the *computer* and the user.

wings: Hidden spaces at the side of a stage, used for entrances and exits.

wipe: (1) A visual *transition* between two video signals using a line or geometric shaped boundary moving across the picture area. **(2)** To erase a recording.

wipe pattern: The geometrically shaped boundary used to perform a *wipe (1)*. Symbols are often used to represent dif   ferent wipe patterns. See *clock wipe, iris wipe* and **Figure W.1**.

wiper: The moving connection on a *potentiometer* control.

wire frame: In *computer graphics*, a three-dimensional image displayed as a series of line segments outlining its surface, including *hidden lines*. See **Figure W.2**.

wireless: Early term for radio transmissions or a radio receiver . (Hist.)

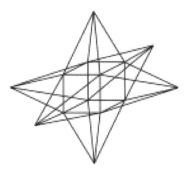

Figure W.2   Wire frame

Wireless Application Protocol (WAP): The *protocol* allowing the *Global System Mobile (GSM)* network to include basic e-mail and *Internet* services.

wireless cable: See *Multichannel Multipoint Distribution System*.

woofer: A *loudspeaker* designed to reproduce only low audio frequencies.

word: The number of *bits* dealt with as a single unit in a digital system.

word clock: A signal used to pass sample rate information between equipment using *BNC* connectors. It is often used as a timing reference in digital audio installations.

workgroup: See *peer-to-peer network*.

work print: (1) In motion picture production, the assembled positive prints of scenes selected by the editor from *rush prints*. Also called *cutting copy*. **(2)** In *videotape* editing, a *videotape* which is created as a result of an *off-line editing* session.

workstation: A computer operating position including the computer, keyboard, monitor and mouse, with perhaps other peripheral equipment including a *scanner (4)*, external data storage, *graphic tablet*, etc. It may also include the desk and other furniture.

World Administrative Radiocommunications Conference (WARC): See *World Radiocommunications Conference*.

World Broadcasting Union (WBU): An organization made up from the world's eight *broadcast* unions: *ABU, ASBU, CBU, EBU, IAB, NABA, OTI* and *URTNA*.

World Radiocommunications Conference (WRC): A four week long conference held every two or three years by the *ITU-R* (see *International Telecommunications Union*) who describe it as the 'forum where countries decide on the shared use of the frequency spectrum to allow the deployment or growth of all types of radiocommunications services, from television and radio broadcasting to mobile telephony , maritime and aeronautical navigation and safety systems and science services '. Was called the *World Administrative Radiocommunication Conference* up to 1992.

Worldwide Television News (WTN): UK television news agency that merged with APTV to form Associated Press Television News (APTN) in 1998. (Hist.)

World Wide Web (WWW): A system of *Internet servers* which supports documents formatted in *HTML* and links to other documents. Invented by Dr Tim Berners-Lee at CERN (Conseil Europ éen pour la Recherche Nucléaire) in Europe.

wow: Periodic variation in speed of an analogue recording medium, occurring less than 10 times per second. See also *flutter.*

wrap: The path of the *magnetic tape* around the *head drum* in a *helical scan* tape recorder.

write: (1) To clock data into a memory. **(2)** To record data on to a magnetic medium.

write once, read many (WORM): A data storage system on which data, once recorded, cannot be altered.

write-on slides: *Slides* with a matt surface, used temporarily during production, which indicate the final image with hand-drawn information.

write protected: Guarding against accidental recording on a magnetic medium, for example a *floppy disk* with its notch covered.

writing speed: (1) In *videotape* systems, the speed at which the spinning *head (2)* crosses the *magnetic tape* surface. It is very much greater than the longitudinal speed of the tape itself. **(2)** In a *cathode ray tube*, the rate at which the electron beam moves across the phosphor while writing.

X

X-axis: In a graph or *oscilloscope*, the horizontal axis of the display .

X-band: The region of the *electromagnetic spectrum* near 8 GHz and used predominantly for *satellite (2)* communications. See **Appendix N.**

xenon lamp: A light source with a compact electrical arc discharge within a quartz envelope containing xenon gas at high pressure.   They are widely used in motion picture film *projectors.*

xerography: A method of producing images by charge patterns on a drum. Particles of dielectric powder (toner) are attracted to the charged surface and transferred onto paper or plastic. They are then fixed on by heat. The system is used in photocopiers and laser printers.

XLR connector: A circular connector with a locking mechanism.   The three-pin version is commonly used in microphones and for other *audio frequency* connections, the four -pin version for 12   *VDC* power connections to portable video and audio equipment. See   **Figure X.1.**

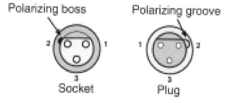

Figure X.1   XLR connector pinout

X-rays: Electromagnetic radiation with wavelengths between those of *gamma rays* and *ultraviolet light.* Discovered in 1895 and so called because their nature was unknown.

X-Y plotter: A paper and ink plotting device where the pen can be driven on the X and *Y axes.* It may be used as a   *computer peripheral.* Also known as a *data plotter.*

Y

Y: Used to indicate the *luminance* signal.

Yagi: A directional *aerial* array invented by Hidetsugu Yagi, a Japanese physicist. It is the form used by the majority of *UHF* television reception *aerials*, comprising a *dipole* connected to the *aerial* lead, a rear reflector and several *directors* in front of the *dipole* whose spacing sets the gain/directivity of the *aerial* across its designed frequency range.

Y-axis: In a graph or *oscilloscope*, the vertical axis of the display .

YCbCr: The digital *luminance* and *colour difference* signals in *ITU-R BT.601* coding. See also *Cb* and *Cr*.

Y/C: Separate analogue *luminance* Y and *chroma* C signals which are recorded on some videotape recording formats. Also the name for the connector carrying the two signals.

Yellow Book: The standard for *CD-ROM*.

YIQ: In the *NTSC* colour television system, the three signals used in the *encoder* to produce the *composite video (2)* output: *luminance* Y and the two *colour difference signals,* I and Q.

yoke: (1) The scanning coils around the neck of a *cathode ray tube* which provide *electromagnetic deflection* of the electron beam(s). Also called *scanning yoke* or *deflection yoke*. **(2)** The *frame (9)* on which a *luminaire* is suspended.

yon plane: In a *computer graphics* display, the back *clipping* plane defining the limit of three-dimensional space furthest from the viewing plane (compare *hither plane*).

YUV: In the *PAL* colour television system, the three signals used in the *encoder* to produce the *composite video (2)* output: *luminance* Y and the two *weighted colour difference signals,* U and V. U is a *weighted* version of the B–Y *colour difference signal* and V is a *weighted* version of the R–Y *colour difference signal*.

Z

Z: Used to indicate *impedance.*

Z-adjustment: Movement perpendicular to the plane of the *X-Y axes.* Specifically in an *overhead projector,* the lamp adjustment to reduce colour fringing when magnification and projection distance are changed.

Z-axis: **(1)** Intensity *modulation* of an *oscilloscope.* **(2)** In *computer graphics,* the third dimension indicating depth.

Z-clipping: In 3D *computer graphics,* limiting the depth of the three dimensional space in front and behind the screen by defining a *hither plane* and a *yon plane,* both parallel to the viewing plane.

zenith: In magnetic tape recording, the extent to which the *head* surface is parallel to the tape. Any error will cause uneven *head* wear and may cause the tape move up or down in an audio tape recorder or poor high frequency recovery in a *videotape recorder.*

zero-crossing: The point on an *AC waveform* where the voltage or current goes through zero as it crosses the time axis.

zero dispersion point: In *fibre optic telecommunications,* the wavelength where *dispersion (3)* is at a minimum. With standard *optical fibre* cable the wavelength is 1310 nm.

zero insertion force (ZIF): A type of socket for multi-pin integrated circuits where the pins are retained or released by lever-operated pressure rather than simple friction. Commonly used with microprocessor chips.

zero level: The standard reference sound level used for lining up equipment. It produces a reading of 4 on a *PPM.*

zero switch: The switch on a *slide projector* for detecting when the *magazine (2)* is off-zero. Also called a homing switch.

zits: Short-term errors on a digital television system. (Colloq.)

zobel network: A phase shifting network sometimes used to ensure stability in *amplifiers* with *negative feedback.*

zones: The division of the area of a video picture *raster* for quality assessment. Generally, zone 1 is the area contained within a circle 0.8 of picture height, zone 2 a circle equal to picture width, and zone 3 is the remaining area outside zone 2. See **Figure Z.1.**

zoom: **(1)** A visual effect which magnifies or shrinks the subject in view. Because the viewpoint does not change, the visual effect on the perspective of the view during a zoom is unnatural. This is a different effect to *tracking (1).* **(2)** In *animation,* moving the camera towards or away from the artwork to photograph larger or smaller fields.

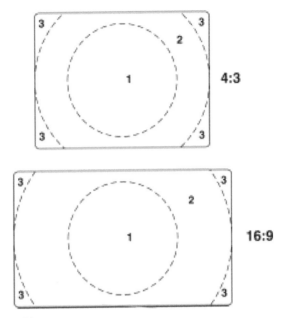

Figure Z.1 Picture zones

zoom lens: A *lens* system for a camera or *projector* whose focal length (magnification) can be continuously varied without loss of *focus (1)*. See also *varifocal lens*.

zoom microphone: A *microphone* whose directional properties can be externally controlled to give a *zoom* effect.

zoom ratio: The ratio of the *focal length* of a lens at the telephoto end of the zoom to that at the wide-angle end. It equals the magnification of the image at one end of the range compared to the other .

Zoopraxinoscope: An early moving image projector . It projected photographs in sequence using a light source, a spinning disc and a lens. It was developed by Jean Louis Meissonier , a nineteenth century French painter. (Hist.)

Zootrope: A viewer using a pair of parallel cylinders revolving on a turntable. Images on the inner disc are viewed through vertical slits in the outer disc to give the illusion of movement. (Hist.)

Numeric

4:1:1 Refers to the ratio of digital video *sampling* frequencies for the *luminance* and two *colour-difference* signals. It is one of a series of ratios defined and specified in *ITU-R BT.601* and is the standard used in the *DVCAM digital videotape recorder* formats.

4:2:0 Refers to the ratio of digital video *sampling* frequencies for the *luminance* and two *colour-difference* signals. It is the standard used in *DVCPRO digital videotape recorder* formats. The colour-difference signals are sampled only on alternate lines.

4:2:2 Refers to the ratio of digital video *sampling* frequencies for the *luminance* and two *colour-difference* signals. It is one of a series of ratios defined and specified in *ITU-R BT.601* and has become the standard used in studio quality digital video systems, *SDI* and the *Digital Betacam, Betacam SX* and *D9 digital videotape recorder* formats.

4:2:2:4 Refers to the ratio of digital video *sampling* frequencies for the *luminance* signal, two *colour-difference* signals and a fourth *key* signal. It is the standard sometimes used in studio quality digital *computer graphics* systems to allow very high quality *keying* operations.

Part 2

Abbreviations and acronyms

A: *ampere*
ABC: **(1)** *American Broadcasting Company.* (Website: www.abc.go.com).
  (2) *Australian Broadcasting Corporation.* (Website: www.abc.net.au).
  (3) *Automatic beam control*
ABR: *Available bit rate*
ABS: *Academy of Broadcast Science* (China)
ABSOC: *Advanced Broadcast Systems of Canada* (Canada)
ABTT: *Association of British Theatr e Technicians* (UK). (Website:
  www.abtt.org.uk)
ABU: Asian-Pacific Broadcasting Union. Part of the *WBU*
AC: *Alternating current*
AC-3: *Audio Code 3*
ACE: **(1)** *Advanced conversion equipment.* **(2)** *American Cinema Editors*
  (USA). (Website: www.ace-filmeditors.org/home.htm)
ACTS: *Advanced communications technology and services*
ACVL: *Association of Cinema and V ideo Laboratories Inc.* (USA)
  (Website: www.acvl.org)
ADA: *Audio distribution amplifier*
ADAC: *Advanced digital adaption converter*
ADC: *Analogue-to-digital converter*
ADO: *Ampex Digital Optics*
ADPCM: *Adaptive differential pulse code modulation*
ADR: *Automated dialogue replacement*
ADSL: *Asymmetric digital subscriber loop*
ADSR: *Attack, decay, sustain, release*
ADT: *Automatic double tracking*
AEEU: *Amalgamated Engineering and Electrical Union* (UK)
AES: *Audio Engineering Society* (USA). (Website: www.aes.org)
AEU: *Amalgamated Engineering Union* (UK). (Hist.)
AF: *Audio frequency*
AFC: *Automatic frequency control*
AFD: *Active format description.*
AGL: *Above ground level*
AFNOR: *Association Française de Normalisation* (France). (Website:
  www.afnor.fr/index_gb)
AFPF: Association Française des Producteurs de Films et de Programmes
  Audiovisuels (France)
AFV: *Audio follow video*
AGC: *Automatic gain control*
AHD: *Audio high density*
AIT: *Advanced intelligent tape (drive)*
ALC: **(1)** *Automatic level control.* **(2)** *Automatic lamp changer*
AM: *Amplitude modulation*

AMPAS: *Academy of Motion Picture Arts and Sciences* (USA)

AMPS: *Association of Motion Pictur e Sound* (UK). (Website: www.amps.net)

ANRS: *Automatic Noise Reduction System*. (Trade name)

ANSI: *American National Standards Institute*. (Website: www.ansi.org)

ANTIOPE: *Acquisition Numérique et T élévisualisation d'Images Organisées en Page d 'Ecriture* (France)

AOR: *Atlantic Ocean region*

API: *Application programme interface*

APL: *Average picture level*

APRS: *Association of Professional Recording Services Ltd* (UK)

ARC: *Aspect ratio converter*

ARD: **(1)** *Arbeitsgemeinschaft der Öffentlichrechtlichen Rundfunkanstalten Deutschlands* (Germany). (Website: www.ard.de). **(2)** *Active region descriptor*

ARPANET: *Advanced Research Projects Agency NETwork* (USA)

ARQ: *Automatic repeat request*

AQL: *Acceptable quality level*

ASA: **(1)** *Advertising Standards Authority* (UK). (Website: www.asa.org.uk). **(2)** *American Standards Association* (USA)

ASBU: Arab States Broadcasting Union. Part of the *WBU*

ASC: *American Society of Cinematographers* (USA)

ASCE: Association of Sound and Communication Engineers (USA)

ASCII: *American Standard Code for Information Inter change*

ASIC: *Applications Specific Integrated Cir cuit*

ASIFA: *Association Internationale du Film d 'Animation*. (Website: www.asifa-hollywood.org)

ASK: *Amplitude shift keying*

ASM: *Assistant stage manager*

AST: *Automatic scan tracking*

ATF: *Automatic track following*

ATM: *Asynchronous transfer mode*

ATRAC: *Adaptive transform acoustic coding*

ATSC: *Advanced Television Systems Committee* (USA). (Website: www.atsc.org)

ATTC: *Advanced Television Technology Center* (USA). (Website: www.attc.org)

ATV: *Advanced Television* (USA)

A/V: *Audio-visual*

AVA: *Audio Visual Association* (UK)

AVC: *Automatic volume control*

AVI: *Audio video interleave*

AVR: *Automatic voltage regulator*

AWG: *Arrayed-waveguide grating*

BABT: *British Approvals Board for Telecommunications*
BAFTA: *British Academy of Film and T elevision Arts* (UK). (Website: www.bafta.org)
BARB: *Broadcasters' Audience Research Board Limited* (UK). (Website: www.barb.co.uk)
BAT: *Bouquet association table*
BBC: *British Broadcasting Corporation.* (Website: www.bbc.co.uk)
BBFC: *British Board of Film Classification* (UK). (Website: www.bbfc.co.uk)
BBS: *Bulletin Board Service*
BCAVM: *British Catalogue of A-V Material* (UK)
BCD: *Binary coded decimal*
BCU: *Big close-up*
BEAB: *British Electrotechnical Approvals Board* (UK). (Website: www.beab.co.uk)
BECTU: *Broadcasting, Entertainment, Cinematograph and Theatr e Union* (UK). (Website: www.bectu.org.uk)
BEAMA: *British Electrotechnical and Allied Manufacturers' Association* (UK). (Website: www.beama.org.uk)
BER: *Bit error rate*
BERT: *Bit error rate test*
BFA: *British Federation of Audio.* (Website: www.british-audio.org.uk)
BFC: *British Film Commission* (UK). (Website: www.britfilmcom.co.uk or www.bfc.co.uk)
BFI: *British Film Institute* (UK). (Website: www.bfi.org.uk)
BIMA: *British Interactive Media Association* (UK). (Website: www.bima.co.uk)
BISFA: *British Industrial and Scientific Film Association.* (See *IVCA*) (Hist.)
BITC: *Burnt in time code*
BKSTS: *British Kinematograph, Sound and T elevision Society* (UK). (Website: www.bksts.com)
B-MAC: *B-type multiplexed analogue component*
BNC: *Bayonet Neil-Concelman*
BP: *Back projection*
BPF: *Band-pass filter*
bps: *Bits per second*
BPSK: *Binary phase shift keying*
BREEMA: *British Radio and Electr onic Equipment Manufactur ers Association*
BRR: *Bit rate reduction*
BS: *British standard* (UK)

BSC: (1) *British Society of Cinematographers Ltd.* (UK). (Website: www.bscine.com). **(2)** *Broadcasting Standards Commission* (UK). (Website: www.bsc.org.uk)

BSI: *British Standards Institution.* (Website: www.bsi.org)

BST: (1) *British Summer T ime* (UK). **(2)** *Bandwidth-segmented transmission*

BT: *British Telecom.* (Website: www.broadcast.bt.com (Broadcast Services))

BTL: *Behind the lens*

BUFVC: *British Universities Film and V ideo Council* (UK). (Website: www.bufvc.ac.uk)

BVA: *British Video Association* (UK). (Website: www.bva.org.uk)

BVU: *Broadcast Video U-matic*

B/W: *Black-and-white*

BWF: *Broadcast wave file*

C: *Coulomb*

CAB: *Canadian Association of Br oadcasters* (Canada). (Website: www.cab-acr.ca)

CABSC: *Canadian Advanced Broadcast Systems Committee* (Canada). (Hist.)

CAD: *Computer aided design*

CAPTAIN: *Character And Pattern T elephone Access Information Net- work (system)* (Japan)

CAR: *Central apparatus room*

CAT: *Conditional access table*

CATV: (1) *Cable Television.* **(2)** *Community Antenna Television*

CAV: (1) *Constant angular velocity.* **(2)** *Component analogue video*

CB: *Citizens Band*

CBA: *Commonwealth Broadcasting Association.* (Website: www.cba.org.uk)

CBR: (1) *Constant bit rate.* **(2)** *Current bit rate*

CBU: Caribbean Broadcasting Union. Part of the *WBU*

CC: *Country Code* (see **Appendix P**)

CCA: *Cable Communications Association* (UK). (Website: www.cable.co.uk)

CCD: *Charge coupled device.*

CCETT: *Centre Commun d'Études de Télédiffusion et Télécommunication* (France). (Website: www.ccett.fr)

CCIR: *Comité Consultatif International de Radiocommunications*

CCITT: *Comité Consultatif International du Télégraphe et du Téléphone*

CCO: *Centre cut out*

CCT: *Circuit*

CCTV: *Closed circuit television*

CCU: (1) *Camera control unit.* **(2)** *Communication concentrator unit*

cd: *candela*

CD: *Compact disc*

CD-DA: *Compact disc – digital audio*

CD-I: *Compact disc – interactive*

CD-R: *Compact disc – recordable*

CD-ROM: *Compact disc – read only memory*

CD-RW: *Compact disc – read write*

CDS: *Cinema Digital Sound.* (Trade name)

CDTV: *Canadian DTV Inc* (Canada)

CD-V: *Compact disc – video*

CEA: *Cinema Exhibitors' Association* (UK)

CED: *Capacitance electronic disc*

CELP: *Codebook excited linear pr ediction*

CEMA: *Consumer Electronics Manufacturers Association.* (Website: www.cemacity.org)

CENELEC: *Comité Européenne de Normalization Electr otechnique* (Website: www.cenelec.be)

CF: *Clean feed*

CG: *Character generator*

CGI: *Computer graphics imaging*

CID: *Compact Iodide, Daylight*. (Trade name)

CIE: *Commission Internationale de l 'Éclairage.* (Website: www.cie.co.at/cie/home.html)

CIF: (1) *Common image format.* **(2)** *Common intermediate format*

CLV: *Constant linear velocity*

C-MAC: *C-type multiplexed analogue component*

CMCR: *Colour mobile control room*

CMRR: *Common mode rejection ratio*

C/N: *Carrier-to-noise ratio*

CNC: Centre Nationale de la Cin ématographie (France)

CNCL: *Commission Nationale de Communications et de la Libert é* (France)

CNR: *Carrier-to-noise ratio*

COFDM: *Coded orthogonal frequency division multiplex*

COM: *Computer output microfilm*

cps: *Characters per second*

CPU: *Central processor unit*

CR: *Carriage return*

CRA: *Camera ready artwork*

CRC: *Cyclic redundancy check*

CRI: *Colour reversal intermediate*

CRT: *Cathode ray tube*
CSDI: *Compressed serial digital interface*
CSI: *Compact Source Iodide.* (Trade name)
CSO: *Colour separation overlay*
CST: Commission Supérieure Technique de l'Image et du Son (France). (Website: www.cst.fr/accueil.html)
CTA: *Cable Television Association* (UK)
CTL: *Control (track)*
CTCM: *Chroma timer compressed multiplex*
CTDM: *Compressed time division multiplex*
CTR: *Cassette tape recorder*
CTRL: *Control (key)*
CU: *Close-up*
CUT: *Circuit under test.* (2) *Co-ordinated Universal Time*
CUTS: *Computer users tape system*

DA: *Distribution amplifier*
DAB: *Digital audio broadcasting*
DAC: *Digital-to-analogue converter*
DAD: *Digital audio disc*
DASH: *Digital audio stationary head*
DAT: *Digital audio tape*
DATV: *Digitally assisted television.* (Hist.)
DAVIC: *Digital Audio Visual Council.* (Website: www.davic.org)
dB: *Decibel*
DBMS: *Database management system*
DBO: *Dead black-out*
DBS: *Direct broadcast satellite*
DC: *Direct Current*
DCMS: *Department for Cultur e, Media and Sports* (UK). (Website: www.culture.gov.uk)
DCT: (1) *Discrete cosine transform.* (2) *Digital component technology*
DDC: *Digital dynamic control*
DDL: *Digital delay line*
DDR: *Digital disk recorder*
DEL: *Direct exchange line*
DGA: *Directors Guild of America* (USA). (Website: www.dga.org)
DICE: *Digital international conversion equipment*
DIF: *Digital interface*
DIGITAG: *Digital Terrestrial Television Action Group.* (Website: www.digitag.org)
DIL: *Dual-in-line*
DIN: *Deutsches Institut für Normung* (Germany)

DMA: *Direct memory access*
D-MAC: *D-type multiplexed analogue component*
DMM: *Digital multi-meter*
DNS: *Domain name service*
DOC: *Drop-out compensator*
DOS: *Disk operating system*
DPCM: *Differential pulse code modulation*
DPE: *Digital production effects.* (Trade name)
DPM: *Digital panel meter*
DRAM: *Dynamic random access memory*
DRAW: *Direct read after write*
D/S: *Downstage*
DSM: *Deputy stage manager*
DSP: *Digital signal processing*
DSS®: *Digital Satellite System.* (Trade name)
DT: *Dynamic tracking*
DTF: *Dynamic track following*
DTG: *Digital Television Group* (UK). (Website: www.dtg.co.uk)
DTH: *Direct to home*
DTI: *Department of Trade and Industry* (UK). (Website: www.dti.gov.uk)
DTP: *Desk-top publishing*
DTRS: *Digital Tape recording system*
DTS: (1) *Digital Theater Systems.* (Website: www.dtstech.com). **(2)** *Decoding time stamp*
DTS-ES: *Digital Theater Systems – Extended Surround*
DTT: *Digital terrestrial television*
DVB: *Digital Video Broadcasting.* (Website: www.dvb.org)
DVD: *Digital versatile disk.*
DVDA: *DVD Association* (Website: www.dvda.org)
DVE: *Digital video effects*
DVI: *Digital video interactive*
DVITC: *Digital vertical interval time code*
DVTR: *Digital video tape recorder*

EAVA: European Audio-Visual Association
EBR: *Electron beam recorder* (Hist.)
EBU: *European Broadcasting Union.* Part of the *WBU.* (Website: www.ebu.ch)
ECC: *Error correction codes*
ECU: *Extreme close-up*
EDL: *Edit decision list*
EDH: *Error detection and handling*
EDTV: *Extended definition television*

EETPU: *Electrical, Electronic, Telecommunications and Plumbing Union* (UK). (Hist.)

EFP: *Electronic field production*

EHT: *Extra-high tension*

EI: *Exposure index*

EIA: *Electronic Industries Association* (USA). (Website: www.eia.org)

EIAJ: *Electronic Industries Association of Japan* . (Website: www.eiaj.or.jp/english/index.htm)

EIRP: *Effective isotropic radiated power*

ELCB: *Earth leakage circuit breaker* (Hist.)

ELF: *Extra low frequency*

ELS: *Extreme long shot*

EMC: *Electromagnetic compatibility*

EMF: *Electromotive force*

EMI: *Electromagnetic interference*

EMM: *Entitlement management message*

ENG: *Electronic news gathering*

EPG: *Electronic programme guide*

EPP: *Electronic post-production*

EPR: *Electronic pin registration*

ES: *Elementary stream*

ETMA: *Educational Television & Media Association* (UK). (Website: www.etma.org.uk)

E-to-E: *Electronic-to-electronic*

ETF: *European Teleconferencing Federation*

ETR: ESTI Technical Report

ETS: *European Telecommunication Standard*

ETSA: European Television Services Association

ETSI: *European Telecommunications Standards Institute.* (Website: www.etsi.org/eds)

EUTELSAT: *European Telecommunications Satellite Or ganization.* (Website: www.eutelsat.com)

FACT: *Federation Against Copyright Theft*

FAP: *Front axial projection*

FAQ: *Frequently asked questions*

FC: *Fibre channel*

FC-AL: *Fibre channel – arbitrated loop*

FCC: (1) *Frame count cueing.* **(2)** *Federal Communications Commission* (USA). (Website: www.fcc.gov/oet)

FDDI: *Fibre distributed data interface*

FEC: *Forward error correction*

FF: *Form feed*
FIAF: *Fédération Internationale des Archives du Film* (France)
FIFO: *First in first out* (memory)
FKGT: *Fernseh-und Kinotechnisehen Geselisehaft*
FM: (1) *Frequency modulation.* (2) *Floor manager*
FOH: *Front of house lights*
FP: *Front projection*
fps: *Frames per second*
FTB: *Fade to black*
FTP: *File transfer protocol*
FX: *Effects*

GBCT: *Guild of British Camera T echnicians* (UK). (Website: www.gbct.org)
GBFE: Guild of British Film Editors (UK)
GEO: *Geostationary Earth orbit*
GHz: *Gigahertz*
GIGO: *Garbage in, garbage out*
GMT: *Greenwich Mean Time* (Hist.)
GOP: *Group of pictures*
GPIB: *General purpose interface bus*
GPS: *Global positioning system*
GSM: *Global system mobile*
GTC: *Guild of Television Cameramen* (UK). (Website: www.gtc.org.uk)
GUI: *Graphical user interface*
G/V: *General view*

H: *Henry*
HD-CIF: *High definition common image format*
HDD: *Hard disk drive*
HD-D5: *High definition – D5*
HD-SDI: *High definition serial digital interface*
HDTV: *High definition television*
HDVS: *High definition video system*
HF: *High frequency*
HFC: *Hybrid fibre-coax*
HPA: *High-power amplifier*
HPF: *High-pass filter*
HTML: *HyperText markup language*
HTTP: *HyperText transport protocol*
Hz: *Hertz*

IAB: **(1)** International Association of Broadcasters. Part of the *WBU*. **(2)** *Internet Architecture Board.* (Website: www.iab.org). **(3)** *International Academy of Broadcasting.* (Website: www.iab.ch)

IABM: *International Association of Broadcasting Manufacturers* (UK). (Website: www.iabm.org.uk)

IAC: *Institute of Amateur Cinematographers* (UK). (Website: www.fvi.org.uk)

IANA: *Internet Assigned Numbers Authority.* (Website: www.iana.org)

IATSE: *International Alliance of Theatrical Stage Employees* (and Moving Picture Machine Operators of the United States and Canada)

IBA: *Independent Broadcasting Authority* (UK) (Hist.)

IBC: *International Broadcasting Convention.* (Website: www.ibc.org)

IBS: *Institute of Broadcast Sound* (UK). (Website: www.ibs.org.uk)

IC: *Integrated circuit*

ICE: **(1)** *Insertion Communication Equipment.* **(2)** *In-circuit emulation*

IDC: *Insulation Displacement Connector*

IEC: *International Electrotechnical Commission.* (Website: www.iec.ch)

IEE: *Institution of Electrical Engineers (UK).* (Website: www.iee.org.uk)

IEEE: *Institute of Electrical and Electronics Engineers, Inc.* (USA). (Website: www.ieee.org)

IETF: *Internet Engineering Task Force.* (Website: www.ietf.org)

IF: *Intermediate film*

IFPP: International Federation of Photogram (and Videogram) Producers

IM: **(1)** *Intermodulation.* **(2)** *Intensity modulation*

IMA: Interactive Multimedia Association (USA)

INMARSAT: *International Marine Satellite Organisation.* (Website: www.inmarsat.org)

INTELSAT: *International Telecommunications Satellite Organisation.* (Website: www.intelsat.int)

IO: *Image orthicon*

IOR: *Indian Ocean region*

IOV: *Institute of Videography* (UK). (Website: www.iov.co.uk)

IP: **(1)** *Internet protocol.* **(2)** *Intermodulation product*

IPA: *Institute of Practitioners in Advertising* (UK). (Website: www.ipa.co.uk)

ips: **(1)** *Inches per second.* **(2)** *Instructions per second*

IR: *Infra-red*

IRD: *Integrated Receiver Decoder*

IRE: *Institute of Radio Engineers* (USA)

IRMA: *International Recording Media Association* (USA). (Website: www.recordingmedia.org)

IRSG: *Internet Research Steering Group*

IRT: *Institut für Rundfunktechnik GmbH* (Germany)

IRTF: *Internet Research Task Force.* (Website: www.irtf.org)
IRTS: *International Radio & T elevision Foundation* (USA). (Website: www.irts.org)
IS: *Internet Society*
ISBA: *Incorporated Society of British Advertisers.* (Website: www.isba.org.uk)
ISDB: *Integrated Services Digital Broadcasting* (Japan)
ISDN: *Integrated Services Digital Network*
ISP: *Internet service provider*
ISTV: *Integrated Services Television* (Japan)
IT: *Information technology*
ITA: **(1)** *International Tape Association* (Hist.) **(2)** *Industrial Telecommunications Association* (USA). (Website: www.ita-relay.com)
ITC: *Independent Television Commission* (UK). (Website: www.itc.org.uk)
ITEA: *International Theatre Equipment Association* (US). (Website: www.itea.com)
ITFC: *Independent Television Facilities Centr e* (UK). (Website: www.itfc.com)
ITN: **(1)** *Independent Television News* (UK). (Website: www.itn.co.uk). **(2)** *Independent Television News* (Sri Lanka).
ITS: **(1)** *Insertion Test Signal.* **(2)** *International Teleproduction Society.* **(3)** Association of Imaging Technology and Sound (US). (W ebsite: www.itsnet.org)
ITU: *International Telecommunications Union.* (Website: www.itu.int)
ITVA: *International Television Association* (USA). (Website: www.itva.org)
ITV: *Independent Television* (UK). (Website: www.itv.co.uk)
IVCA: *International Visual Communications Association.* (Website: www.ivca.org)

J: *joule*
JEDEC: *Joint Electronic Devices Engineering Council*
JPEG: *Joint Photographic Expert Gr oup.* (Website: www.jpeg.org)

K: *Kelvin*
kHz: *Kilohertz*
kVA: *Kilovolt ampere*
kW: *Kilowatt*

L: *Lambert*
LAD: *Laboratory aim density*
LAN: *Local area network*

LCD: *Liquid crystal display*
LED: *Light-emitting diode*
LEO: *Low Earth orbit*
LF: **(1)** *Line feed.* **(2)** *Low frequency*
LHCP: *Left-hand circular polarization*
LINX: *London Internet Exchange* (Website: www.linx.net)
lm: *Lumen*
LOR: *Laser optical reflection*
LOT: *Laser optical transmission*
LPF: *Low-pass filter*
LS: **(1)** *Loudspeaker.* **(2)** *Long shot*
LSB: *Least significant bit*
LTC: *Longitudinal time code*
lx: *Lux*

MAC: **(1)** *Multiplexed analogue component.* **(2)** *Media access control*
MADI: *Multi-channel audio digital interface*
MATV: *Master antenna television*
MCB: *Miniature circuit breaker*
MCPC: *Multiple channel per carrier*
MCPS: *Mechanical Copyright Protection Society (UK)*
MCR: **(1)** *Master Control Room.* **(2)** *Mobile Control Room*
MCS: *Medium close shot*
MCU: **(1)** *Medium close-up.* **(2)** *Multipoint conference unit*
MD: *MiniDisc*
MDS: *Multipoint distribution system*
M/E: *Mix/effects*
ME: *Metal evaporated*
MEO: *Medium Earth orbit*
MF: *Medium frequency*
MHP: *Multimedia home platform*
MHz: *Megahertz*
MIDI: *Musical Instrument Digital Interface*
mips: *Million instructions per second*
MLS: *Medium long shot*
MMDS: **(1)** *Multichannel multipoint distribution system* **(2)** *Microwave multipoint distribution system*
MOD: **(1)** *Minimum object distance.* **(2)** *Magneto optical disk*
MOSAIC: *Methods for Optimisation and Subjective Assessment in Image Communications*
MP: *Metal particle*
MPAA: *Motion Picture Association of America* (USA). (Website: www.mpaa.org)

MPEAA: *Motion Picture Export Association of America* (USA)
MPEG: (1) *Moving Picture Experts Group.* (Website: www.cselt.it/mpeg/).
(2) *Motion Picture Editors Guild* (Website: www.editorsguild.com/)
MPSE: *Motion Picture Sound Editors* . (Website: www.mpse.org/index.html)
MPU: *Microprocessor unit*
MS: *Medium shot*
MSB: *Most significant bit*
MSO: *Multiple system operator*
MTBF: *Mean time between failures*
MTF: *Modulation transfer function*
MUSA: (1) *Maximum use of surface area.* (2) *Multiple unit steerable antenna*
MUSE: *Multiple sub-Nyquist sampling encoding*
MUSICAM: *Masking pattern adapted Universal Sub-band Integrated Coding And Multiplexing*

N: *Newton*
NAB: *National Association of Broadcasters* (USA). (Website: www.nab.org)
NABA: *North American Broadcasters Association.* Part of the *WBU* (Website: www.nabanet.com)
NABT: *North American Broadcast Teletext* (USA)
NARTB: *National Association of Radio and Television Broadcasters* (USA). (Hist.)
NATO: *National Association of Theatre Owners* (USA). (Website: www.hollywood.com/nato/)
NBS: *National Bureau of Standards* (USA). (Hist.)
NC: *Normally closed*
NCR: Network control room
NCTA: *National Cable Television Association* (USA). (Website: www.ncta.com/home)
ND: *Neutral density (filter)*
NFS: *Network file system*
NICAM: *Near Instantaneously Companded Audio Multiplex*
NIST: *National Institute of Standards and Technology* (USA)
NIT: *Network information table*
NLQ: *Near letterpress quality*
NMPFT: *National Museum of Photography, Film & Television* (Bradford, UK). (Website: www.nmpft.org.uk/home.asp)
NO: *Normally open*
NPL: *National Physical Laboratory* (UK). (Website: www.npl.co.uk)
NRZ: *Non-return to zero*
NTC: *Negative temperature coefficient*

NTL: *National Transcommunications Limited* (UK). (Website: www.ntl.com

NTP: *Normal temperature and pressure*

NTSC: *National Television Systems Committee*

OB: *Outside broadcast*

OCR: *Optical character recognition*

Oe: *Oersted*

OEM: *Original equipment manufacturer*

OFDM: *Orthogonal frequency division multiplex*

OHP: *Over-head projector*

OIRT: *Organisation Internationale de Radiodiffusion et T élévision* (Hist.)

OMFI: *Open media framework interchange*

OOV: *Out of vision*

OSD: *On screen display*

OSI: *Open system interconnect*

OSS: *Over the shoulder shot*

OTDR: *Optical time domain reflectometer*

OTF: *Optical transfer function*

OTI: Organisacion de la Television Iberoamericana. Part of the *WBU*

OTS: *Orbital test satellite*. (Hist.)

Pa: *Pascal*

PA: **(1)** *Public address.* **(2)** *Production assistant.* **(3)** *Power amplifier* (US). **(4)** *Press Association* (UK). (Website: www1.pa.press.net)

PACT: *Producers Alliance for Cinema and T elevision.* (Website: www.pact.co.uk)

PAL: **(1)** *Phase alternating line.* **(2)** *Programmable array logic*

PAT: *Programme association table*

PC: *Personal computer*

PCB: *Printed circuit board*

PCM: *Pulse code modulation*

PCMCIA: *Personal Computer Memory Card Interface Association*

PCR: *Programme clock reference*

PD: *Potential difference*

PDA: **(1)** *Pulse distribution amplifier.* **(2)** *Post deflection acceleration*

PEC: *Photo-electric cell*

PFL: *Pre-fade listen*

PID: *Programme identification number*

PIP: *Picture in picture*

PLL: *Phase locked loop*

PLUGE: *Picture line-up generating equipment*

PMT: (1) *Programme map table.* (2) *Photo mechanical transfer*
PoP: *Point of presence*
POR: *Pacific Ocean region*
p-p: *Peak-to-peak*
ppm: *Parts per million*
PPM: (1) *Peak programme meter.* (2) *Pulse position modulation*
pps: *Pictures per second*
PPT: Projected Picture Trust (UK)
PPV: *Pay per view*
PRS: *Performing right society.* (Website: www.prs.co.uk)
PS: *Program stream*
PSC: *Portable single camera*
PSI: (1) *Programme specific information.* (2) *Pounds per square inch*
PSK: *Phase-shift keying*
PTC: (1) *Positive temperature coefficient.* (2) *Piece to camera*
PTH: *Plated through holes*
PTS: *Presentation time stamp*
PTT: *Post, telegraph and telephone*
PU: *Pick-up*
PVR: *Personal video recorder*
PWM: *Pulse-width modulation*

QAM: *Quadrature amplitude modulation*
QCIF: *Quarter common intermediate format*
QEF: *Quasi error-free*
QIL: *Quad-in-line*
QPSK: *Quadrature phase-shift keying*

RA: *Radiocommunications Agency* (UK)
RAB: *RAID Advisory Board* (Website: www.raid-advisory.com)
RACE: *Research into Advanced Communications for Europe.* (Hist.)
RAID: *Redundant array of inexpensive disks*
RAM: *Random access memory*
RCD: *Residual current device*
RDS: *Radio data service*
RF: *Radio frequency*
RFC: *Request for comments document*
RFI: *Radio frequency interference*
RGB: *Red green blue*
RHCP: *Right-hand circular polarization*
RIAA: *Recording Industry Association of America* (Website: www.riaa.com)
RKO: *Radio-Keith-Orpheum* (USA). (Website: www.rko.com)
RMS: *Root mean square*

RO: *Receive only*
ROM: *Read only memory*
RPS: *Royal Photographic Society* (UK). (Website: www.rps.org)
RTA: *Real time analyser*
RTS: *Royal Television Society* (UK). (Website: www.rts.org.uk)
RTZ: *Return to zero*
RX: (1) *Receiver.* **(2)** *Re-transmit*
RZ: *Return to zero*

S: *Siemens*
SAG: *Safe area generator*
SBCA: *Satellite Broadcasting and Communications Association* (USA). (Website: www.sbca.com)
SBE: *Society of Broadcast Engineers Inc.* (USA). (Website: www.sbe.org)
SCA: *Subsidiary Communications Authorisation*
SCART: *Syndicat des Constructeurs d 'Appareils Radiorécepteurs et Téléviseurs*
SCH: *Subcarrier to horizontal*
SCPC: *Single channel per carrier*
SCR: (1) *Silicon controlled rectifier.* **(2)** *System clock reference*
SCSI: *Small computer system interface*
SDDI: *Serial digital data interface*
SDDS: *Sony Dynamic Digital Sound.* (Website: www.sdds.com)
SDH: *Synchronous digital hierarchy*
SDI: *Serial digital interface*
SDTI: *Serial data transport interface*
SECAM: *Sequential Couleur à Mémoire*
SES: *Société Européene des Satellites.* (Website: www.ses-astra.com)
SFD: Society of Film Distributors (UK)
SFN: *Single frequency network*
SFX: (1) *Sound effects.* **(2)** *Special effects*
SHF: *Super high frequency*
SI: *Système International*
SIF: *Source intermediate format*
SIL: *Single-in-line*
SLP: *Super long play*
SMATV: *Satellite Master Antenna Television*
SMD: *Surface mount device*
SMG: *Scottish Media Group*
SMPTE: *Society of Motion Pictur e and Television Engineers* (Website: www.smpte.org)
S/N: *Signal-to-noise ratio*
SNG: *Satellite news gathering*

SNR: *Signal-to-noise ratio*
SNV: *Satellite news vehicle*
SOF: *Sound-on-film*
SPDIF: *Sony/Philips Digital InterFace*
SPG: *Sync-pulse generator*
SPL: *Sound pressure level*
SQIF: *Sub-quarter common intermediate format*
SRAM: *Static random access memory*
SSVR: *Solid state video recorder*
STB: *Set-top box*
STP: *Shielded twisted pair*
S-VHS: *Super-VHS*

T: *Tesla*
TARIF: *Technical apparatus for the rectification of indifferent film*
TBC: *Time-base corrector*
TCIP: *Time code in picture*
TCP: *Transmission control protocol*
TCP/IP: *Transmission control protocol/Internet protocol*
TDM: *Time division multiplex*
TEA: *Theatre Equipment Association* (USA). (Hist.)
TES: *Transportable Earth Station*
TFT: *Thin film transistor*
THD: **(1)** *Total harmonic distortion.* **(2)** *Third harmonic distortion*
TIA: *Telecommunications Industries Association.* (Website: www.tia.org)
TIM: *Transient intermodulation distortion*
TK: *Telecine*
TMD: *Thermal magnetic duplication*
TS: *Transport stream*
TTL: **(1)** *Through the lens.* **(2)** *Transistor-transistor logic*
TV: *Television*
TVRO: *Television receive only*
TVS: *Television South* (Hist.)
TWT: *Travelling wave tube*
TX: *Transmission*

UBR: *Uncommitted bit rate*
UDP: *User datagram protocol*
UHF: *Ultra high frequency*
UI: *Unit interval*
UMTS: *Universal Mobile Telecommunications System*
UNM: *United News & Media* (UK). (Website: www.unm.co.uk)
URL: *Uniform (or universal or unique) resource locator*

Abbreviations and acronyms

URTNA: Union des Radiodiffusions et Television Nationales d'Afrique.
  Part of the *WBU*
UTC: *Universal Time Co-ordinated*
UV: *Ultra-violet*

V: *Volt*
VA: *Volt-ampere*
VAC: *Volts, alternating current*
VAR: *Vision apparatus room*
VBI: *Vertical blanking interval*
VBR: *Variable bit rate*
VCA: **(1)** *Voltage controlled attenuator.* **(2)** *Voltage controlled amplifier*
VCC: *Video Compact Cassette* (Hist.)
VCO: *Voltage Controlled Oscillator*
VCPS: *Video Copyright Protection Society* (UK)
VCR: *Video cassette recorder*
VDA: **(1)** *Video distribution amplifier.* **(2)** *Video Dealers' Association* (UK)
VDC: *Volts, direct current*
VDT: *Visual display terminal* (Hist.)
VDU: *Visual display unit*
VESA: *Video Electronics Standards Association.* (Website: www.vesa.org)
VHD: *Video high density*
VHF: *Very high frequency*
VHS: *Video home system*
VIPS: *Visual information processing system*
VITC: *Vertical interval time code*
VITS: *Vertical insertion test signal*
VLC: *Variable length coding*
VLF: *Very low frequency*
VLP: *Very long play*
VO: *Voice over*
VOM: *Volt-ohm meter*
VSB: *Vestigial sideband*
VT: *Videotape*
VTR: *Video tape recorder*
VTVM: *Vacuum tube volt meter*
VU: *Volume unit*

W: *Watt*
WAP: *Wireless Application Protocol*
WARC: *World Administrative Radiocommunication Conference.* (Hist.)
Wb: *Weber*
WBU: *World Broadcasting Union*

WDM: *Wavelength division multiplexing*
WIMP: *Windows, icons, mouse and pull-down menus*
WORM: *Write once, read many*
WRC: *World Radiocommunications Conference*
WSS: *Wide screen signalling*
WTN: *Worldwide Television News* (UK). (Hist.)
WWW: *World Wide Web*
WYSIWYG: *What you see is what you get*

Y: *Luminance*

ZIF: *Zero insertion force*

Part 3

Appendices

Appendix A
Image areas for motion picture films

Table App.A.1 Image areas in film

Film gauge	Camera aperture (mm)		Maximum projected area (mm)	
	Width	Height	Width	Height
Super-8 mm	5.69	4.22	5.46	4.01
16 mm	10.05	7.42	9.65	7.26
Super-16 (Type W)	12.52	7.42	–	–
35 mm Academy	21.95	16.05	21.11	15.29
35 mm Widescreen 1.85:1	21.95	12.00	21.11	11.41
35 mm Anamorphic	21.95	18.80	21.29	18.21
65 mm/70 mm	52.50	23.00	48.59	22.10
IMAX	70.00	52.50	–	–

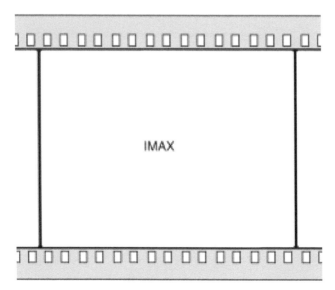

Figure App.A.1   Imax image area

Appendix A: Image areas for motion picture films

The relevant SMPTE standards are: SMPTE 7, SMPTE 59, SMPTE 152, SMPTE 195, SMPTE 201M, SMPTE 215 and SMPTE 233. See **Appendix U**.

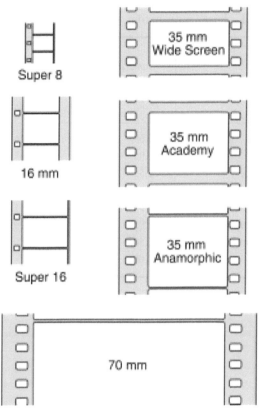

Figure App.A.2 Film image areas

Appendix B
Image areas used in television

Table App.B.1   Film image areas for 4:3 television

	Transmitted area (mm)	Safe action area (mm)	Safe title area (mm)	Corner radius (mm)
16 mm film	9.35 × 7.00	8.4 × 6.3	7.5 × 5.6	1.5
35 mm film	20.12 × 15.10	18.1 × 13.6	16.1 × 12.1	3.2
50 × 50 mm slides	28.60 × 21.50	25.7 × 19.3	22.8 × 17.2	4.6

For 4:3 television, the dimensions of the safe action area and safe title area represent 90 per cent and 80 per cent respectively of the transmitted area.

For 16:9 television, the safe action area height is 93 per cent of the picture height and the safe title area is 90 per cent.

Note that the change to display screens with little or no rounded corners means that safe areas are now usually taken to be fully rectangular .

Full details of cinematography standards are given in British Standard BS 5550, which is divided into the following sub-sections:

BS 5550:1	8 mm and Super-8
BS 5550:2	16 mm
BS 5550:3	35 mm
BS 5550:4	65 mm and 70 mm
BS 5550:5	Common to several gauges
BS 5550:6	Television usage
BS 5550:7	Production and presentation

Appendix B: Image areas used in television

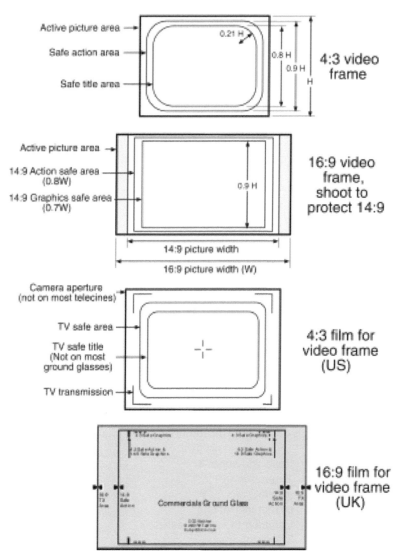

Figure App.B.1   Image areas in television

Appendix C
Image areas for slides

Table App.C.1   Image areas for slides

	Width (mm)	Height (mm)
Superslide (46 mm film in 50 × 50 mm mount)		
Camera area	39.1	39.1
Mount aperture	37.5	37.5
35 mm slide (35 mm film in 50 × 50 mm mount)		
Camera area	36.0	24.0
Mount aperture	34.3	22.5
Half-frame mount aperture	22.5	15.5
Miniature slide (110 format film in 30 × 30 mm mount)		
Mount aperture	17.0	13.0
35 mm film strip		
Camera area	24.0	18.0
Mount aperture	22.35	16.75

The relevant SMPTE standard is SMPTE 96M (see **Appendix U**).

Appendix D
Picture aspect ratios

Table App.D.1   Picture aspect ratios

Format	Aspect ratio	Application
Video	1.33:1 (4:3)	Television, 4:3 display
Super-8 film	1.33:1	Amateur film (Hist.)
16 mm film	1.33:1	Film and television 4:3 presentation
Imax 70 mm film	1.33:1	Very large screen presentation
Film-strip	1.33:1	Slide projection
Film-strip	1.5:1	Slide projection
Video	1.56:1 (14:9)	Television, widescreen display presentation in letterbox or in pillarbox on 4:3 display
35 mm film	1.65:1	Widescreen presentation
Super 16 mm film	1.67:1 (15:9)	Widescreen presentation
Video	1.78:1 (16:9)	Television, 16:9 widescreen display
35 mm film	1.85:1	Widescreen presentation
70 mm film	2.2:1	Widescreen presentation
35 mm film	2.35:1	Anamorphic projection
35 mm film × 3	2:7:1	Cinerama presentation (Hist.)

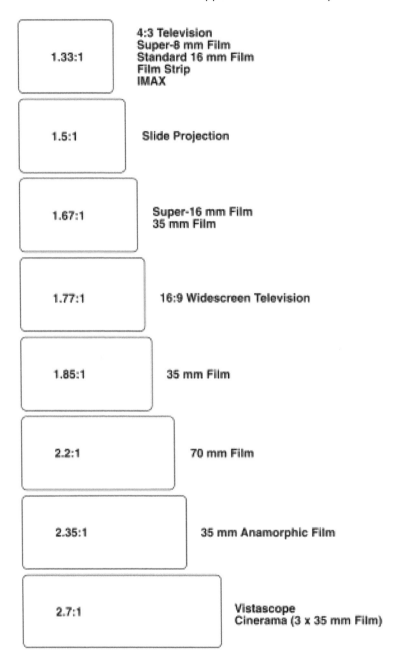

Figure App.D.1 Picture aspect ratios

Appendix E
Projected image size calculation

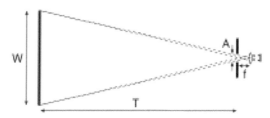

Figure App.E.1 Projected image size

Normal projection

The width of the projected picture, $W = (T \times A)/f$

Anamorphic projection

The width of the projected picture, $W = (T \times A \times 2)/f$

where,

W = projected picture width (m)
T = projection distance (m)
A = projector aperture width (mm). (See Table App.E.1.)
f = focal length of projection lens (mm)

Table App.E.1   Projector aperture widths

Film type	Aperture width
Super-8 film	5.4 mm
16 mm film	9.6 mm
35 mm film	21.1 mm
Film strip and half-slide	22.5 mm
Standard 50 mm slide	34.5 mm

Appendix F
Recommended seating plan

In general, the seating layout should have the front row not closer to the screen than two screen-widths, with the back row not more than six screen-widths away. Seats should preferably be kept within a wedge area of about 20 degrees from each side of the screen. This can be extended to 30 degrees for non-directional matt white surfaces. For more directional screen materials it is desirable to keep within 15 degrees.

Screen luminance and other viewing conditions are defined in standard SMPTE 196M (see **Appendix U**).

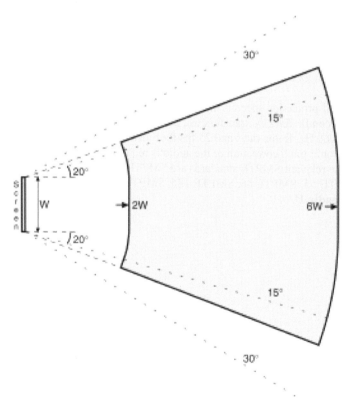

Figure App.F.1   Recommended seating plan

Appendix G
Sound track position and displacement on film

In motion picture prints, the separation between any picture frame and its corresponding sound is standardized, with a tolerance of $\pm 1/2$ frame.

Super-8 optical track: sound 22 frames ahead of picture
16 mm optical track: sound 26 frames ahead of picture
35 mm optical track: sound 20 frames ahead of picture

Super-8 magnetic track: sound 18 frames ahead of picture
16 mm magnetic track: sound 28 frames ahead of picture
35 mm magnetic track: sound 28 frames behind picture
70 mm magnetic track: sound 24 frames behind picture

35 mm *SDDS* track: 0 frames ahead of picture
35 mm *DTS* track: Not applicable, sound on separate CD-ROM
35 mm *Dolby Digital* track: 26 frames ahead of picture

Normal projection speed is internationally standardized at 24 fps (frames per second). *Telecine* transfers use 24 fps for television systems based on 59.94/60 Hz frame rates and 25 fps for those based on 50 Hz, when electronic pitch correction of the audio is required.

The relevant SMPTE standards are SMPTE 40, SMPTE 41, SMPTE 48, SMPTE 55, SMPTE 86, SMPTE 185, SMPTE 203 and SMPTE 221. See **Appendix U**.

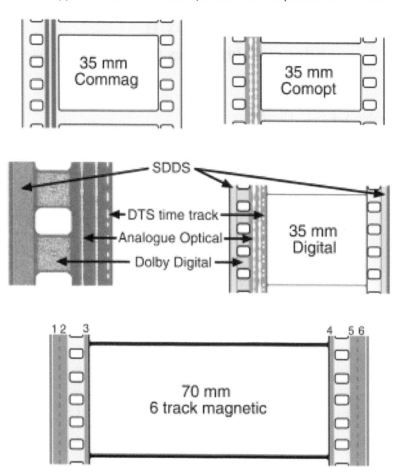

Figure App.G.1 Sound track locations on film prints

Appendix H
Film running times and footage

Table App.H.1   Film running times

	35 mm @ 24 fps		16 mm @ 24 fps		16 mm @ 25 fps	
Mins	Feet	Metres	Feet	Metres	Feet	Metres
1	90	27.4	36	11.0	37.5	11.4
2	180	54.9	72	22.0	75.0	22.9
3	270	82.3	108	32.9	112.5	34.3
4	360	109.7	144	43.9	150.0	45.7
5	450	137.2	180	54.9	187.5	57.2
6	540	164.6	216	65.9	225.0	68.6
7	630	192.0	252	76.8	262.5	80.0
8	720	219.5	288	87.8	300.0	91.4
9	810	246.9	324	98.8	337.5	102.9
10	900	274.3	360	109.7	375.0	114.3
15	1350	411.5	540	164.6	562.5	174.5
20	1800	548.7	720	219.5	750.0	228.6
25	2250	685.8	900	274.3	937.5	285.8
30	2700	823.0	1080	329.2	1125.0	342.9
35	3150	960.1	1260	384.1	1312.5	400.5
40	3600	1097.3	1440	438.9	1500.0	457.2
45	4050	1234.5	1620	493.8	1687.5	514.4
50	4500	1371.6	1800	548.7	1875.0	571.5
55	4950	1508.8	1980	603.5	2062.5	628.7
60	5400	1654.9	2160	658.4	2250.0	685.8
120	10800	3291.9	4320	1316.8	4500	1371.6
180	16200	4937.8	6480	1975.1	6750	2057.4

Appendix I
World television standards

At the time of writing (June 2000) many countries had still to decide on which digital television standards to adopt, so the digital information is provisional. The various systems for analogue and digital television are summarized in **Appendix R** and **Appendix S**.

Table App.I.1 World analogue and digital television standards

| Country | Analogue | | Digital | | |
	Colour	System	Satellite	Cable	Terrestrial
Afghanistan	PAL	D	DVB-S	–	–
Albania	SECAM	B			
Algeria	PAL	B	DVB-S	DVB-C	DVB-T
Andorra	PAL	B			
Angola	PAL	I			
Antigua and Barbuda	NTSC	M			
Antilles (The Netherlands)	NTSC	M			
Argentina	PAL	N	DVB-S	DVB-C	–
Armenia	SECAM	D/K			
Aruba	NTSC	M			
Ascension Island	PAL	I			
Australia	PAL	B	DVB-S	DVB-C	DVB-T
Austria	PAL	B/G	DVB-S	DVB-C	DVB-T
Azerbaijan	SECAM	D/K			
Azores	PAL	B			
Bahamas	NTSC	M			
Bahrain	PAL	B	DVB-S	–	–
Bangladesh	PAL	B	DVB-S	–	–

Table App.I.1 (*Continued*)

Country	Analogue		Digital		
	Colour	System	Satellite	Cable	Terrestrial
Barbados	NTSC	N			
Belarus	SECAM	D/K	DVB-S	DVB-C	–
Belgium	PAL	B/G	DVB-S	DVB-C	DVB-T
Belize	NTSC	M			
Benin	SECAM	K			
Bermuda	NTSC	M			
Bhutan	PAL	B	DVB-S	–	–
Bolivia	NTSC	N	DVB-S	–	–
Bosnia and Herzegovina	PAL	B	DVB-S	DVB-C	DVB-T
Botswana	PAL	I	DVB-S	–	–
Brazil	PAL-M	M	DVB-S	DVB-C	–
Brunei Darussalam	PAL	B			
Bulgaria	SECAM	D	DVB-S	DVB-C	DVB-T
Burundi	SECAM	K			
Cambodia	PAL	B			
Cameroon	PAL	B			
Canada	NTSC	M	DVB-S/ DSS	DVB-C/ OpenCable	ATSC
Canary Islands	PAL	B			
Cape Verde	PAL	I			
Central African Republic	PAL	B	DVB-S	–	–
Chad	SECAM	K			
Chile	NTSC	M	DVB-S	–	–
China	PAL	D	DVB-S	DVB-C	–
Colombia	NTSC	M	DVB-S	–	–
Comoros	PAL	I			
Congo	SECAM	D	DVB-S	–	–

Table App.I.1 *(Continued)*

| Country | Analogue | | Digital | | |
	Colour	System	Satellite	Cable	Terrestrial
Costa Rica	NTSC	M			
Côte d'Ivoire (Ivory Coast)	SECAM	K	DVB-S	–	–
Croatia	PAL	B	DVB-S	DVB-C	DVB-T
Cuba	NTSC	M	DVB-S	DVB-C	–
Cyprus	PAL	B			
Czech Republic	SECAM	D	DVB-S	DVB-C	DVB-T
Denmark	PAL	B	DVB-S	DVB-C	DVB-T
Djibouti	SECAM	K			
Dominica	NTSC	M			
Dominican Republic	NTSC	M			
Ecuador	NTSC	M	DVB-S	–	–
Egypt	SECAM	B	DVB-S	DVB-C	DVB-T
El Salvador	NTSC	M	DVB-S	–	–
Estonia	PAL/ SECAM	D/K	DVB-S	DVB-C	–
Ethiopia	PAL	B	DVB-S	–	–
Falkland Islands	PAL	I			
Fiji	NTSC	M			
Finland	PAL	B/G	DVB-S	DVB-C	DVB-T
France	SECAM	L	DVB-S	DVB-C	DVB-T
Gabon	SECAM	K			
Gambia	PAL	I	DVB-S	–	–
Georgia	SECAM	D/K			
Germany	PAL	B/G	DVB-S	DVB-C	DVB-T
Ghana	PAL	B	DVB-S	–	–
Gibraltar	PAL	B			
Greece	PAL	B	DVB-S		DVB-T

Table App.I.1 *(Continued)*

Country	Analogue		Digital		
	Colour	System	Satellite	Cable	Terrestrial
Greenland	NTSC	M	DVB-S	–	DVB-T
Grenada	NTSC	M			
Guadeloupe	SECAM	K			
Guam	NTSC	M			
Guatemala	NTSC	M			
Guinea	PAL	B			
Guinea Bissau	PAL	I			
Guyana	SECAM	K			
Hawaii	NTSC	M			
Honduras	NTSC	M			
Hong Kong	PAL	I	DVB-S	–	–
Hungary	SECAM	K	DVB-S	DVB-C	DVB-T
Iceland	PAL	B	DVB-S	DVB-C	DVB-T
India	PAL	B	DVB-S	DVB-C	DVB-T
Indonesia	PAL	B	DVB-S	–	–
Iran	SECAM	B	DVB-S	–	–
Iraq	SECAM	B	DVB-S	–	–
Ireland	PAL	I	DVB-S	DVB-C	DVB-T
Israel	PAL	B/G	DVB-S	–	DVB-T
Italy	PAL	B/G	DVB-S	DVB-C	DVB-T
Jamaica	NTSC	M			
Japan	NTSC	M	DVB-S	DVB-C	ISDB-T
Jordan	PAL	B	DVB-S	–	–
Kazakhstan	SECAM	D/K	DVB-S	DVB-C	–
Kenya	PAL	B	DVB-S	–	–
Korea (North)	PAL/ NTSC	D	DVB-S	–	–
Korea (South)	NTSC	M	DVB-S/ DSS	DVB-C/ OpenCable	–

Table App.I.1 (*Continued*)

Country	Analogue		Digital		
	Colour	System	Satellite	Cable	Terrestrial
Kuwait	PAL	B	DVB-S	–	–
Kyrgyzstan	SECAM	D/K			
Latvia	SECAM	D/K	DVB-S	DVB-C	–
Lebanon	SECAM	B	DVB-S	–	–
Lesotho	PAL	I	DVB-S	–	–
Liberia	PAL	B			
Libya	PAL	B	DVB-S	DVB-C	–
Liechtenstein	PAL	B			
Lithuania	SECAM	D/K	DVB-S	DVB-C	–
Luxembourg	PAL/ SECAM	C/L	DVB-S	DVB-C	DVB-T
Macau	PAL	I			
Macedonia (former Yugoslav Republic)	PAL	B	DVB-S	DVB-C	DVB-T
Madagascar	SECAM	K			
Madeira	PAL	B			
Malawi	PAL	B			
Malaysia	PAL	B	DVB-S	DVB-C	DVB-T
Mali	SECAM	K			
Malta	PAL	B			
Martinique	SECAM	K			
Mauritania	SECAM	K			
Mauritius	SECAM	B			
Mexico	NTSC	M	DVB-S	–	–
Moldova	SECAM	D/K	DVB-S	DVB-C	–
Monaco	PAL/ SECAM	C/G/L	DVB-S	DVB-C	DVB-T
Mongolia	SECAM	D	DVB-S	–	–
Morocco	SECAM	K	DVB-S	DVB-C	DVB-T

Table App.I.1 (*Continued*)

| Country | Analogue | | Digital | | |
	Colour	System	Satellite	Cable	Terrestrial
Mozambique	PAL	I	DVB-S	–	–
Myanmar (Burma)	NTSC	N	DVB-S	–	–
Namibia	PAL	B	DVB-S	–	–
Nepal	PAL	B	DVB-S	–	–
Netherlands	PAL	B/G	DVB-S	DVB-C	DVB-T
New Caledonia	SECAM	K			
New Zealand	PAL	B	DVB-S	DVB-C	DVB-T
Nicaragua	NTSC	M			
Niger, Republic of	SECAM	K	DVB-S	–	DVB-T
Nigeria	PAL	B	DVB-S	–	–
Norway	PAL	B/G	DVB-S	DVB-C	DVB-T
Oman	PAL	B/G	DVB-S	–	–
Pakistan	PAL	B	DVB-S	–	–
Panama	NTSC	M	DVB-S	–	–
Paraguay	PAL	N	DVB-S	–	–
Peru	NTSC	M	DVB-S	–	–
Philippines	NTSC	M	DVB-S	–	–
Poland	SECAM	D/K	DVB-S	–	DVB-T
Polynesia (French)	SECAM	K			
Portugal	PAL	B/G	DVB-S	DVB-C	DVB-T
Puerto Rico	NTSC	M			
Qatar	PAL	B	DVB-S	–	–
Reunion	SECAM	K			
Romania	SECAM	D	DVB-S	DVB-C	DVB-T
Russian Federation	SECAM	D/K	DVB-S	DVB-C	DVB-T
Rwanda	SECAM	K	DVB-S	–	–
St. Helena	PAL	I			
St. Kitts & Nevis	NTSC	M			

Table App.I.1 (*Continued*)

Country	Analogue		Digital		
	Colour	System	Satellite	Cable	Terrestrial
St. Pierre and Miquelon	SECAM	K			
St. Vincent and Grenadines	NTSC	M			
Saint Lucia	NTSC	M			
Samoa, American	NTSC	M			
Samoa, Western	PAL	B			
Saudi Arabia	PAL	B	DVB-S	–	–
Senegal	SECAM	L			
Seychelles	PAL	I			
Singapore	PAL	B	–	–	DVB-T
Slovakia (Slovak Republic)	PAL/ SECAM	B/K	DVB-S	DVB-C	DVB-T
Slovenia	PAL	B	DVB-S	DVB-C	DVB-T
Somalia	PAL	B/G			
South Africa	PAL	I	DVB-S	DVB-C	DVB-T
Spain	PAL	B	DVB-S	DVB-C	DVB-T
Sri Lanka	PAL	B	DVB-S	–	–
Sudan	PAL	B			
Suriname	NTSC	M			
Swaziland	PAL	I	DVB-S	–	–
Sweden	PAL	B/G	DVB-S	DVB-C	DVB-T
Switzerland	PAL	B/G	DVB-S	DVB-C	DVB-T
Syria	PAL	B	DVB-S	–	–
Tajikistan	SECAM	D/K			
Taiwan	NTSC	M			
Tanzania	PAL	B			
Thailand	PAL	B	DVB-S	–	–
Togo	SECAM	K			
Tonga	NTSC	M			

Table App.I.1 (*Continued*)

Country	Analogue		Digital		
	Colour	System	Satellite	Cable	Terrestrial
Trinidad and Tobago	NTSC	M			
Tunisia	SECAM	B	DVB-S	DVB-C	–
Turkey	PAL	B	DVB-S	DVB-C	DVB-T
Turkmenistan	SECAM	D/K	DVB-S	DVB-C	–
Turks and Caicos Islands	NTSC	M			
Uganda	PAL	B	DVB-S	–	–
Ukraine	SECAM	D/K	DVB-S	DVB-C	–
United Arab Emirates	PAL	B	DVB-S	–	–
United Kingdom	PAL	I	DVB-S	DVB-C	DVB-T
United States of America	NTSC	M	DVB-S/ DSS	DVB-C/ OpenCable	ATSC
Uruguay	PAL	N	DVB-S	–	–
Uzbekistan	SECAM	D/K			
Vatican City State	PAL	B/G			
Venezuela	NTSC	M	DVB-S	–	–
Vietnam	SECAM/ NTSC	D/M			
Virgin Islands, British	NTSC	M			
Virgin Islands (USA)	NTSC	M			
Yemen	PAL/ NTSC	B	DVB-S	–	–
Yugoslavia (Serbia, Montenegro)	PAL/ SECAM	B/K	DVB-S	DVB-C	DVB-T
Zaire	SECAM	K	DVB-S	–	–
Zambia	PAL	B	DVB-S	–	–
Zimbabwe	PAL	B	DVB-S	–	–

Appendix J
Analogue and digital videotape formats

Table App.J.1 Comparison of some analogue videotape formats

	1″ C	Betacam SP	U-Matic Hi-band	Hi-8	S-VHS	VHS
Originator	Ampex/Sony	Sony	Sony	Sony	JVC	JVC
First shown	1978	1987	1977	1990	1987	1977
Tape width (mm)	25.4	12.65	19	8	12.65	12.65
Tape speed (cm/s)	24	10.15	9.5	2	2.34	2.34
Tape thickness (µm)	28	13	27	13 (MP) 10 (ME)	21	21
Tape type*	MO	MP	MO	MP, ME	MO	MO
Video write speed (cm/s)	2140	575	854	312	485	485
Head drum diameter (mm)	134.6	74.5	110	40	62	62
Head gap azimuth differences	0°	± 15°	0°	±8°	± 6°	± 6°
Video track width (µm)	160	86 (lum.) 73 (col.)	85	34.4	58	58

* MO = metal oxide, MP = metal particle, ME = metal evaporated

Table App.J.2 Comparison of some digital videotape formats

	Digital Betacam	Betacam SX	D9 (Digital-S)	DVCPRO	DVCAM	DV	D-VHS
Originator	Sony	Sony	JVC	Panasonic	Sony	All	JVC
First shown	1994	1996	1996	1995	1996	1995	1998
Recorded data standard	Sony Intra-field	MPEG-2 4:2:2P@ML	DVC Intra-frame	DVC Intra-frame	DVC Intra-frame	DVC Intra-frame	MPEG-2 MP@ML, MP@HL
Bit-rate (Mbit/s)	126	18	50	25	25	25	28.2 (high-def. mode) 14.2 (standard mode) 4.7 (low-speed mode)
Tape width (mm)	12.65	12.65	12.65	6.35	6.35	6.35	12.65
Tape speed (cm/s)	9.7	5.96	5.78	3.38	2.82	1.88	Various
Tape type*	MP	MP	MP	MP	ME	MP, ME	MO
Head drum diameter (mm)	74.5	74.5	21.7	21.7	21.7	21.7	62
Data compression ratio	2.2:1	10:1	3.3:1	5:1	5:1	5:1	Various
Compatibility	Replays Betacam SP tapes	Replays Betacam SP tapes	Replays S-VHS tapes	Replays DV and DVCAM tapes	Replays DV tapes	–	Replays VHS, S-VHS tapes
Video track width (μm)	26	32	20	18	15	10	

*MO = metal oxide, MP = metal particle, ME = metal evaporated

Appendix K
Analogue television line-rate waveforms

The drawings show the waveforms of the line blanking periods for a 625-line 50-field television system and a 525-line 59.94-field system.

Notes:

1 The 525-line NTSC waveform for studio use is defined in standard SMPTE 170M (see **Appendix U**).
2 In North America, line blanking duration in 525-lines was originally specified at 20 IRE above blanking level, but is now measured at 4 IRE (50 per cent of set-up).
3 In North America, front porch and sync to set-up durations in 525-lines are measured from 4 IRE above blanking level to leading edge of sync.
4 No pedestal is used in Japan.
5 The video amplitude for digital 525-line systems is 700 mV, not 714 mV (140 IRE) and there is no pedestal.

Table App.K.1 Signal parameters for 625-line and 525-line video signals

Parameter	625-line/50 Hz	525-line/59.94 Hz
Line period (μs)	64	63.56
Line blanking (μs)	12.05	10.9
Front porch (μs)	1.55	1.5
Sync pulse (μs)	4.7	4.7
Back porch (μs)	–	4.7
Burst start (μs)	5.6	5.3
Sync to set-up (μs)	–	9.4
Sync amplitude	300 mV	40 IRE (286 mV)
White level	700 mV	100 IRE (714 mV)
Pedestal (set-up)	0 mV	7.5 IRE (54 mV)
Burst amplitude	300 mV	40 IRE (286 mV)

Appendix K: Analogue television line-rate waveforms

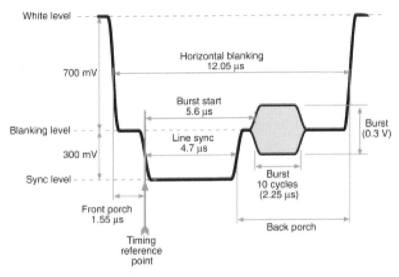

Figure App.K.1 625-line horizontal blanking period

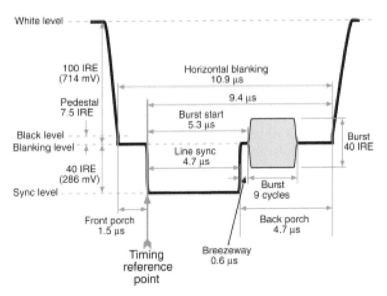

Figure App.K.2 525-line horizontal blanking period

Appendix L
Analogue television field blanking periods

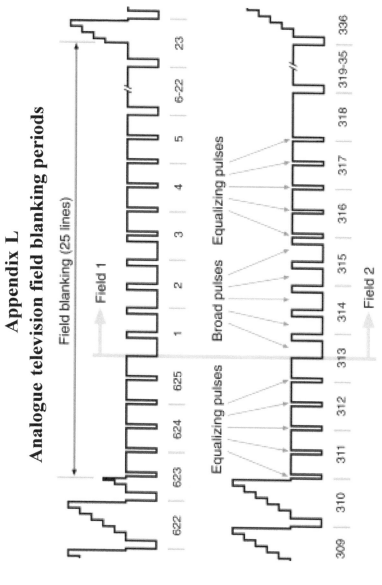

Figure App.L.1 625-line vertical blanking interval

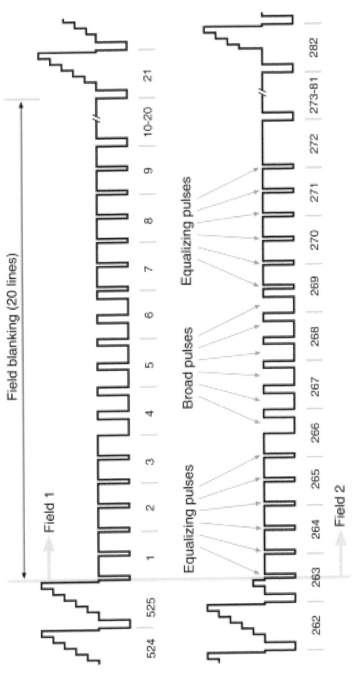

Figure App.L.2 525-line vertical blanking interval

Appendix M
SI units and prefixes for numerical factors

Table App.M.1   SI units

Basic SI Units			Some Derived SI Units		
Name	*Symbol*	*Quantity*	*Name*	*Symbol*	*Quantity*
ampere	A	Electric current	coulomb	C	Electric charge
candela	cd	Luminous intensity	farad	F	Capacitance
			hertz	Hz	Frequency
kelvin	K	Temperature	henry	H	Inductance
kilogram	kg	Mass	joule	J	Work, energy, heat
metre	m	Length	lumen	lm	Luminous flux
second	s	Time	lux	lx	Illumination
			newton	N	Force
			ohm	Ω	Electrical resistance
			pascal	Pa	Pressure
			radian	rad	Angle
			siemens	S	Electrical conductivity
			steradian	sr	Solid angle
			tesla	T	Magnetic flux density
			volt	V	Electromotive force
			watt	W	Power
			weber	Wb	Magnetic flux

Appendix M: SI units and prefixes for numerical factors

Table App.M.2 Numerical factors

Factor	Prefix	Symbol	Multiplier
10^{24}	yotta	Y	1 000 000 000 000 000 000 000 000
10^{21}	zetta	Z	1 000 000 000 000 000 000 000
10^{18}	exa	E	1 000 000 000 000 000 000
10^{15}	peta	P	1 000 000 000 000 000
10^{12}	tera	T	1 000 000 000 000
10^{9}	giga	G	1 000 000 000
10^{6}	mega	M	1 000 000
10^{3}	kilo	k	1000
10^{2}	hecto	h	100
10	deca	da	10
10^{-1}	deci	d	0.1
10^{-2}	centi	c	0.01
10^{-3}	milli	m	0.001
10^{-6}	micro	μ	0.000 001
10^{-9}	nano	n	0.000 000 001
10^{-12}	pico	p	0.000 000 000 001
10^{-15}	femto	f	0.000 000 000 000 001
10^{-18}	atto	a	0.000 000 000 000 000 001
10^{-21}	zepto	z	0.000 000 000 000 000 000 001
10^{-24}	yocto	y	0.000 000 000 000 000 000 000 001

Appendix N
The electromagnetic spectrum

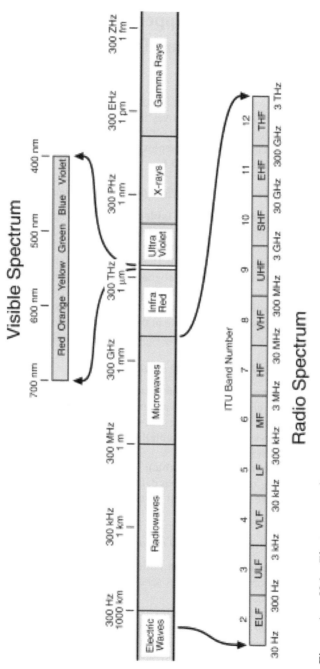

Figure App.N.1 Electromagnetic spectrum

Appendix O
MPEG-2 and DVB data tables

In addition to carrying packets representing video and audio, an MPEG-2 *transport stream (TS)* must convey information allowing the user and their receiver to access the available services/programmes. This data is in the form of a set of tables. The complex process of decoding a TS programme is illustrated in **Figure App.O.1**.

MPEG-2 PSI (program specific information) mandatory tables

- **PAT (program allocation table):** Carried in packets with PID = 0. Links allocated programme number (0 –65 535) with PID (*packet identifier*) of packets carrying the PMT for each TS programme.
- **PMT (program map table):** There is one PMT for each TS programme, indicating the PIDs of the programme' s *elementary streams* and the *program clock reference (PCR)* needed to synchronize the decoder .
- **CAT (conditional access table):** Carried in packets with PID = 1. Present when at least one programme in the TS has *conditional access*. Indicates the PIDs of *entitlement management message (EMM)* packets.

DVB-SI (service information) mandatory tables

- **NIT (network information table):** Carries information (e.g. frequencies, channel numbers) about a network of more than one TS. It is programme number 0 in the TS.
- **SDT (service description table):** Lists the names and other information about each service in the TS.
- **EIT (event information table):** Carries information on *events (2)* happening or about to happen in the current (or perhaps another) TS.
- **TDT (time and date table):** Used to update the *real time* clock in the receiver.

DVB-SI (service information) optional tables

- **BAT (bouquet association table):** Groups services together for presentation to the viewer. A service or programme can be part of more than one *bouquet*.
- **RST (running status table):** Update information for a change in status of an *event (2)*, only transmitted when the status changes.
- **ST (stuffing tables):** A table of meaningless information used as a space filler to maintain the required bit-rate.

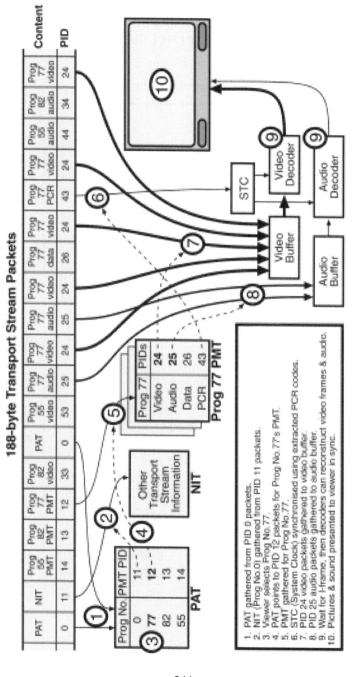

Figure App.O.1   DVB decoding

Appendix P
ITU, ISO and Internet country codes

Table App.P.1   ITU, ISO and Internet country codes

Country	Country Code			
	ITU	*ISO-2*	*ISO-3*	*Internet*
Afghanistan	AFG	AF	AFG	.af
Alaska	ALS	–	–	–
Albania	ALB	AL	ALB	.al
Algeria	ALG	DZ	DZA	.dz
Andorra	AND	AD	AND	.ad
Angola	AGL	AO	AGO	.ao
Anguilla	ANG	AI	AIA	.ai
Antarctica	ANT	AQ	ATA	.aq
Antigua and Barbuda	ATG	AG	ATG	.ag
Antilles (The Netherlands)	ATN	AN	ANT	.an
Argentina	ARG	AR	ARG	.ar
Armenia	ARM	AM	ARM	.am
Aruba	ARU	AW	ABW	.aw
Ascension Island	ASC	–	–	.ac
Australia	AUS	AU	AUS	.au
Austria	AUT	AT	AUT	.at
Azerbaijan	AZE	AZ	AZE	.az
Azores	AZR	–	–	–
Bahamas	BAH	BS	BHS	.bs
Bahrain	BHR	BH	BHR	.bh
Bangladesh	BGD	BD	BGD	.bd
Barbados	BRB	BB	BRB	.bb
Belarus	BLR	BY	BLR	.by
Belgium	BEL	BE	BEL	.be

Table App.P.1 *(Continued)*

Country	Country Code			
	ITU	*ISO-2*	*ISO-3*	*Internet*
Belize	BLZ	BZ	BLZ	.bz
Benin	BEN	BJ	BEN	.bj
Bermuda		BM	BMU	.bm
Bhutan	BHU	BT	BTN	.bt
Bolivia	BOL	BO	BOL	.bo
Bosnia and Herzegovina	BIH	BA	BIH	.ba
Botswana	BOT	BW	BWA	.bw
Brazil	B	BR	BRA	.br
Brunei Darussalam	BRU	BN	BRN	.bn
Bulgaria	BUL	BG	BGR	.bg
Burkina Faso	BFA	BF	BFA	.bf
Burundi	BDI	BI	BDI	.bi
Cambodia	CBG	KH	KHM	.kh
Cameroon	CME	CM	CMR	.cm
Canada	CAN	CA	CAN	.ca
Canary Islands	CNR	–	–	–
Cape Verde	CPV	CV	CPV	.cv
Caroline Islands (Palau)	CAR	PW	PLW	.pw
Cayman Islands	CYM	KY	CYM	.ky
Central African Republic	CAF	CF	CAF	.cf
Chad	TCD	TD	TCD	.td
Chile	CHL	CL	CHL	.cl
China	CHN	CN	CHN	.cn
Christmas Islands	CHR	CX	CXR	.cx
Cocos-Keeling Islands	ICO	CC	CCK	.cc
Colombia	CLM	CO	COL	.co
Comoros	COM	KM	COM	.km
Congo, Republic of	COG	CG	COG	.cg
Cook Islands	CKH	CK	COK	.ck
Costa Rica	CTR	CR	CRI	.cr

Table App.P.1 *(Continued)*

Country	Country Code			
	ITU	*ISO-2*	*ISO-3*	*Internet*
Côte d'Ivoire (Ivory Coast)	CTI	CI	CIV	.ci
Croatia	HRV	HR	HRV	.hr
Cuba	CUB	CU	CUB	.cu
Cyprus	CYP	CY	CYP	.cy
Czech Republic	CZE	CZ	CZE	.cz
Denmark	DNK	DK	DNK	.dk
Djibouti	DJI	DJ	DJI	.dj
Dominica	DMA	DM	DMA	.dm
Dominican Republic	DOM	DO	DOM	.do
Ecuador	EQA	EC	ECU	.ec
Egypt	EGY	EG	EGY	.eg
El Salvador	SLV	SV	SLV	.sv
Equatorial Guinea	GNE	GQ	GNQ	.gq
Estonia	EST	EE	EST	.ee
Ethiopia	ETH	ET	ETH	.et
Faeroe Islands	FRI	FO	FRO	.fo
Falkland Islands	FLK	FK	FLK	.fk
Fiji	FJI	FJ	FJI	.fj
Finland	FNL	FI	FIN	.fi
France	F	FR	FRA	.fr
French Guiana	GUF	GF	GUF	.gf
Gabon	GAB	GA	GAB	.ga
Gambia	GMB	GM	GMB	.gm
Georgia	GEO	GE	GEO	.ge
Germany	D	DE	DEU	.de
Ghana	GHA	GH	GHA	.gh
Gibraltar	GIB	GI	GIB	.gi
Great Britain	G	–	–	–
Greece	GRC	GR	GRE	.gr
Greenland	GRL	GL	GRL	.gl

Table App.P.1 (*Continued*)

Country	ITU	ISO-2	ISO-3	Internet
		Country Code		
Grenada	GRD	GD	GRD	.gd
Guadeloupe	GDL	GP	GLP	.gp
Guam	GUM	GU	GUM	.gu
Guatemala	GTM	GT	GTM	.gt
Guinea	GUI	GN	GIN	.gn
Guinea-Bissau	GNB	GW	GNB	.gw
Guyana	GUY	GY	GUY	.gy
Haiti	HTI	HT	HTI	.ht
Hawaii	HWA	–	–	–
Honduras	HND	HN	HND	.hn
Hong Kong	HKG	HK	HKG	.hk
Hungary	HNG	HU	HUN	.hu
Iceland	ISL	IS	ISL	.is
India	IND	IN	IND	.in
Indonesia	INS	ID	IDN	.id
Iran	IRN	IR	IRN	.ir
Iraq	IRQ	IQ	IRQ	.iq
Ireland	IRL	IE	IRL	.ie
Israel	ISR	IL	ISR	.il
Italy	I	IT	ITA	.it
Jamaica	JMC	JM	JAM	.jm
Japan	J	JP	JPN	.jp
Jordan	JOR	JO	JOR	.jo
Kazakhstan	KAZ	KZ	KAZ	.kz
Kenya	KEN	KE	KEN	.ke
Kiribati	KIR	KI	KIR	.ki
Korea (North)	KRE	KP	PRK	.kp
Korea (South)	KOR	KR	KOR	.kr
Kuwait	KWT	KW	KWT	.kw
Kyrgyzstan	KGZ	KG	KGZ	.kg

Table App.P.1   (*Continued*)

Country	Country Code			
	ITU	*ISO-2*	*ISO-3*	*Internet*
Lao Peoples Democratic Republic	LAO	LA	LAO	.la
Latvia	LVA	LV	LVA	.lv
Lebanon	LBN	LB	LBN	.lb
Lesotho	LSO	LS	LSO	.ls
Liberia	LBR	LR	LBR	.lr
Libyan Arab Jamahiriya	LBY	LY	LBY	.ly
Liechtenstein		LI	LIE	.li
Lithuania	LTU	LT	LTU	.lt
Luxembourg	LUX	LU	LUX	.lu
Macau		MO	MAC	.mo
Macedonia (former Yugoslav Republic)	MDN	MK	MKD	.mk
Madagascar	MDG	MG	MDG	.mg
Madeira	MDR	–	–	–
Malawi	MWI	MW	MWI	.mw
Malaysia	MLA	MY	MYS	.my
Maldives	MLD	MV	MDV	.mv
Mali	MLI	ML	MLI	.ml
Malta	MLT	MT	MLT	.mt
Mariana, Northern	MRA	MP	MNP	.mp
Martinique	MRT	MQ	MTQ	.mq
Mauritania	MTN	MR	MRT	.mr
Mauritius	MAU	MU	MUS	.mu
Mayotte	MYT	YT	MYT	.yt
Mexico	MEX	MX	MEX	.mx
Moldova, Republic of	MDA	MD	MDA	.md
Monaco	MCO	MC	MCO	.mc
Mongolia	MNG	MN	MNG	.mn
Montserrat	MSR	MS	MSR	.ms
Morocco	MRC	MA	MAR	.ma
Mozambique	MOZ	MZ	MOZ	.mz

Table App.P.1 *(Continued)*

Country	Country Code			
	ITU	*ISO-2*	*ISO-3*	*Internet*
Myanmar (Burma)	BRM	MM	MMR	.mm
Namibia	NMB	NA	NAM	.na
Nauru	NRU	NR	MRU	.nr
Nepal	NPL	NP	NPL	.np
Netherlands	HOL	NL	NLD	.nl
New Caledonia	NCL	NC	NCL	.nc
New Zealand	NZL	NZ	NZL	.nz
Nicaragua	NCG	NI	NIC	.ni
Niger, Republic of	NGR	NE	NER	.ne
Nigeria	NIG	NG	NGA	.ng
Niue Islands	NIU	NU	NIU	.nu
Norway	NOR	NO	NOR	.no
Oman	OMA	OM	OMN	.om
Pakistan	PAK	PK	PAK	.pk
Panama	PNR	PA	PAN	.pa
Panama Canal Zone	PNZ	–	–	–
Papua New Guinea	PNG	PG	PNG	.pg
Paraguay	PRG	PY	PRY	.py
Peru	PRU	PE	PER	.pe
Philippines	PHL	PH	PHL	.ph
Poland	POL	PL	POL	.pl
Polynesia (French)	OCE	PF	PYF	.pf
Portugal	POR	PT	PRT	.pt
Puerto Rico	PTR	PR	PRI	.pr
Qatar	QAT	QA	QAT	.qa
Réunion Island	REU	RE	REU	.re
Romania	ROU	RO	ROM	.ro
Russian Federation	RUS	RU	RUS	.ru
Rwanda	RRW	RW	RWA	.rw
St. Helena	SHN	SH	SHN	.sh

Table App.P.1 *(Continued)*

Country	Country Code			
	ITU	*ISO-2*	*ISO-3*	*Internet*
St. Kitts & Nevis	SCN	KN	KNA	.kn
Saint Lucia	LCA	LC	LCA	.lc
St. Pierre and Miquelon	SPM	PM	SPM	.pm
St. Vincent and Grenadines	VCT	VC	VCT	.vc
Samoa, American	SMA	AS	ASM	.as
Samoa, Western	SMO	WS	WSM	.ws
São Tome e Principe	STP	ST	STP	.st
Saudi Arabia	ARS	SA	SAU	.sa
Senegal	SEN	SN	SEN	.sn
Seychelles	SEY	SC	SYC	.sc
Singapore	SNG	SG	SGP	.sg
Slovakia (Slovak Republic)		SK	SVK	.sk
Slovenia	SVN	SI	SVN	.si
Solomon Islands	SLM	SB	SLB	.sb
Somalia	SOM	SO	SOM	.so
South Africa	AFS	ZA	ZAF	.za
Spain	E	ES	ESP	.es
Sri Lanka	CLN	LK	LKA	.lk
Sudan	SDN	SD	SDN	.sd
Suriname	SUR	SR	SUR	.sr
Swaziland	SWZ	SZ	SWZ	.sz
Sweden	S	SE	SWE	.se
Switzerland	SUI	CH	CHE	.ch
Syrian Arab Republic	SYR	SY	SYR	.sy
Tajikistan	TJK	TJ	TJK	.tj
Taiwan	TWN	TW	TWN	.tw
Tanzania	TZA	TZ	TZA	.tz
Thailand	THA	TH	THA	.th
Togo	TGO	TG	TGO	.tg

Table App.P.1 (*Continued*)

Country	Country Code			
	ITU	*ISO-2*	*ISO-3*	*Internet*
Tonga	TON	TO	TON	.to
Trinidad and Tobago	TRD	TT	TTO	.tt
Tristan da Cunha	TRC	–	–	–
Tunisia	TUN	TN	TUN	.tn
Turkey	TUR	TR	TUR	.tr
Turkmenistan	TKM	TM	TKM	.tm
Turks and Caicos Islands	TCA	TC	TCA	.tc
Tuvalu	TUV	TV	TUV	.tv
Uganda	UGA	UG	UGA	.ug
Ukraine	UKR	UA	UKR	.ua
United Arab Emirates	UAE	AE	ARE	.ae
United Kingdom	G	GB	GBR	.uk
United States	USA	US	USA	.us
Uruguay	URG	UY	URY	.uy
USSR (former, now obsolete)	URS	–	–	.su
Uzbekistan	UZB	UZ	UZB	.uz
Vanuatu	VUT	VU	VUT	.vu
Vatican City State (Holy See)	CVA	VA	VAT	.va
Venezuela	VEN	VE	VEN	.ve
Vietnam	VTN	VN	VNM	.vn
Virgin Islands, British	VRG	VG	VGB	.vg
Virgin Islands (USA)	VIR	VI	VIR	.vi
Wallis and Futuna Islands	WAL	WF	WLF	.wf
Yemen	YEM	YE	YEM	.ye
Yugoslavia (Serbia, Montenegro)	YUG	YU	YUG	.yu
Zaire	ZAI	ZR	ZAR	.zr
Zambia	ZMB	ZM	ZMB	.zm
Zimbabwe	ZWE	ZW	ZWE	.zw

Appendix Q
World television broadcasters

A1TV   America One Television (USA). (Website: www.americaone.com)
A.2   Antenne 2 (France). (Website: www.vtcom.fr)
ABBS   Antigua and Barbuda Broadcasting Services (Antigua and Barbuda)
ABC   (1) American Broadcasting Company (USA). (Website: www.abc.go.com)
   (2) Australian Broadcasting Corporation (Australia). (Website: www.abc.net.au)
AIR   All India Radio (India)
AMC   American Movie Channel (USA). (Website: www.amctv.com)
APTN   Associated Press Television News (UK). (Website: www.aptn.com/aptn/copystore.nsf/pages/aptn+homepage)
APTV   Alabama Public Television (USA)
ARD   Arbeitsgemeinschaft der Öffentlichrechtlichen Rundfunkanstalten Deutschlands (Germany). (Website: www.ard.de)
ART   Arab Radio and Television (Saudi Arabia). (Website: www.art-tv.net)
ATA   Asociacion de Teleradiodifusoras Argentinas (Argentina)
ATC   Argentina Televisora Color (Argentina)
ATV   Asia Television Limited (Hong Kong). (Website: www.hkstar.com/atv)
AVRO   Algemene Vereniging Radio Omroep (Holland). (Website: www.omroep.nl/avro)
BAS   Broadcasting Authority of Swaziland (Swaziland)
BBC   (1) Bermuda Broadcasting Company Limited (Bermuda)
   (2) British Broadcasting Corporation (UK). (Website: www.bbc.co.uk)
BBS   Brunei Broadcasting Service (Brunei)
BCC   Broadcasting Corporation of China (Formosa)
BCTV   British Columbia Television (Canada)
BFBS   British Forces Broadcasting Service (UK). (Website: www.ssvc.com/bfbs/index.htm)
BiB   British Interactive Broadcasting (UK). (Website: www.bib.co.uk)

BNT	Bulgarian National Television (Bulgaria)
BRT	Belgische Radio en Televisie (Belgium, Flemish language)
BRTN	Belgische Radio en Televisie Nederlands (Belgium, Dutch language). (Website: www.brtn.be)
BSkyB	British Sky Broadcasting (UK). (Website: www.sky.com/home)
BSN	British Satellite News (UK)
BTV	(1) Bahrain Television (Bahrain)
	(2) Beijing Television (China)
	(3) Botswana Television (Botswana)
C4	Channel 4 (UK). (Website: www.channel4.com)
C5	Channel 5 (UK). (Website: www.channel5.co.uk)
CBC	(1) Canadian Broadcasting Corporation (Canada). (Website: www.cbc.ca)
	(2) Caribbean Broadcasting Corporation (Barbados)
	(3) Cologne Broadcasting Centre (Germany)
	(4) Cyprus Broadcasting Corporation (Cyprus)
CBS	Columbia Broadcasting System (USA). (Website: www.cbs.com)
CCTV	China Central Television (China)
CBSat	China Broadcasting Film & Television Satellite Company (China)
CFI	Canal France International (France)
CMT	Country Music Television (USA)
CNN	Cable News Network (USA). (Website: www.cnn.com)
CNBC	Joint business television service of NBC & Dow Jones (UK). (Website: www.cnbc.co.uk)
CRTV	Cameroon Radio-Television Corp. (Cameroon)
CSD	Canal Satellite Digital (France)
CST	Ceskoslovenska Televize (Czech Republic). (Website: www.czech-tv.cz)
CTG	Cisneros Television Group (Venezuela)
CTN	Chinese Television Network (Hong Kong). (Website: www.ctn.net)
CTS	Chinese Television System (Taiwan). (Website: www.cts.com.tw)
CTV	(1) Canadian Television network (Canada). (Website: www.ctv.ca)
	(2) Channel Television (UK). (Website: www.channeltv.co.uk)
CVN	Caribbean Video Satellite Television (Trinidad & Tobago)
DBC	Digital Broadcast Centre (Egypt)
DBL	Defontes Broadcasting Limited (Bermuda)

DD	Doordarshan (India)
DRTV	Danmarks Radio (Denmark). (Website: www.dr.dk)
DSF	Deutsches Sport Fernsehen (Germany)
DW	Deutsche Welle (South Africa-radio)
EDTV	Emirates Dubai Television (UAE)
EIRT	Ethnikon Idryma Radiophonias-Tileoraseos (Greece)
ENT	Estonia News Television (Estonia). (Website: www.utv.ee)
ERT	Elliniki Radiophonia-Tikleorassi (Greece)
ESC1	Egyptian Satellite Channel 1 (Egypt)
ETC	Eesti Televisiooni (Estonia)
ETV	Ethiopian Television (Ethiopia)
FBC	(1) Fiji Broadcasting Commission (Fiji)
	(2) Fox Broadcasting Company (USA). (Website: www.foxnetwork.com)
FR2	France Région 2 (France). (Website: www.france2.fr)
FR3	France Région 3 (France). (Website: www.france3.fr)
FTC	Fuji Telecasting Company (Japan)
FTN	Fuji Television Network (Japan)
GBC	(1) Ghana Broadcasting Corporation (Ghana)
	(2) Gibraltar Broadcasting Corporation (Gibraltar)
GBT	Great Belize Television (Belize)
GEU	Granada Entertainment USA (USA)
GMG	Granada Media Group (UK). (Website: www.granada.co.uk)
GTV	Granada Television (UK). (Website: www.granada.co.uk)
HBO	Home Box Office (USA). (Website: www.hbo.com)
HRT	Croatia. (Website: www.hrt.hr)
HTV	Harlech Television (UK). (Website: www.htv.co.uk)
IBC	International Broadcasting Corporation (Thailand)
IBTE	Iraqi Broadcasting and Television Establishment (Iraq)
ICR	Institute Cubano de Radiodifusion (Cuba)
IFTN	Ireland Film & Television Network (Eire). (Website: www.iftn.ie)
INBS	Icelandic National Broadcasting Service (Iceland). (Website: www.ruv.is)
INRTI	Instituto Nacional de Radio y Television Intravision (Congo)
IRIB	Islamic Republic of Iran Televison (Iran). (Website: www.irna.com)
IRT	Institut fur Rundfunktechnik GmbH (Germany)
IRTV	Iranian Television (Iran). (Website: www.irtv.com)
IRV	Insituto Nacional de Radio y Television (Colombia)
ITN	(1) Independent Television News (UK). (Website: www.itn.co.uk)
	(2) Independent Television Network (Sri Lanka)

ITV	(1) Independent Television (UK). (Website: www.itv.co.uk) (2) Independent Television (Canada). (Website: www.itv.ca/index.htm)
ITV2	Independent Television 2. The second ITV channel (UK). (Website: www.itv2.co.uk)
JBC	Jamaica Broadcasting Corporation (Jamaica)
JRT	Jugoslevenska Radiotelevizija (or Udruzenje Jugoslovenskil Radiotelevizija d.O.O.) (Yugoslavia)
JRTV	Jordan Radio & Television Corporation. (Website: www.jrtv.com/index.html)
JSC	Jazeera Satellite Channel (Qatar)
JTV	Jordan Television Corporation (Jordan)
KBC	Kenya Broadcasting Corporation (Kenya)
KBS	Korean Broadcasting System (Korea). (W ebsite: www.kbs.co.kr)
KBTS	Kuwait Broadcasting and Television Service (Kuwait)
KET	Kentucky Educational Television (USA)
KNR	Kalaallit Nunaata Radioa (Greenland)
KRO	Katholieke Radio Omroep (Holland). (W ebsite: www.omroep.nl/kro)
KTN	Kenya Television Network (Kenya)
LBC	London Broadcasting Company (UK). (W ebsite: www.lbc.co.uk)
LBCI	Lebanese Broadcasting Corporation International (Lebanon). (Website: www.inco.com.lb/lbci.com.lb)
LNN	London News Network (UK). (W ebsite: www.lnn-tv.co.uk)
LNT	Latvijas Neatkariga Televizija (Latvia)
LTV	Latvijas Televizija (Latvia)
LVRTC	Latvijas Valsts Radio un Televizijas Centrs (Latvia)
LWT	London Weekend Television (UK). (Website: www.lwt.co.uk)
MBC	(1) Middle East Broadcast Centre (Middle East) (2) Munhwa Broadcasting Corporation (Korea). (W ebsite: www.mbc.co.kr)
MBS	(1) Mauritius Broadcasting Service (Mauritius) (2) Mutual Broadcasting System (USA – radio)
MCOT	Mass Communication Organization of Thailand (Thailand). (Website: www.tv3.co.th)
MTV	(1) Magyar Televizio (Hungary) (2) Macedonia Television (Former Republic of Macedonia) (3) Marpin-TV (Dominica) (4) Mersey Television (UK). (Website: www.merseytv.com) (5) Music Television (USA). (Website: www.mtv.com)

MTV *cont.*	(6) Oy Mainos-TV-Reklam Ab (Finland). (Website: www.mtv3.fi)
	(7) Murr Television (Lebanon). (Website: www.mtv.com.lb)
MUTV	Manchester United Television (UK). (Website: www.manutd.com)
NBC	(1) National Broadcasting Commission (Nigeria)
	(2) National Broadcasting Company (USA). (Website: www.nbc.com)
NBN	Nagoya Broadcasting Network (Japan). (Website: www.nbn.co.jp/nbn_home/nbn.html)
NCRV	Nederlandse Christelijke Radio Vereniging (Holland). (Website: www.omroep.nl/ncrv)
NET	Nippon Educational Television (Japan)
NHK	Nippon Hoso Kyokai (Japan). (Website: www.nhk.or.jp/index-e.html)
NOB	Nederlands Omroepproduktie Bedrijf (Holland)
NOS	Nederlandisehe Omroep Stichting (or Nederlandse Omroepprogramma Stichting) (Holland)
NPR	National Public Radio (USA). (Website: www.npr.org)
NPS	Nederlandse Programma Stichting (Holland). (Website: www.omroep.nl/nps)
NRK	Norsk Rikskringkasting (Norway). (Website: www.nrk.no)
NRU	Nederlandse Radio-Unie (Holland)
NSB	Nordic Satellite Broadcasting (based in Brussels, Belgium)
NTS	Nederlandse Televisie Stichting (Holland)
NTV	(1) Nippon Television Network Corporation (Japan). (Website: www.ntv.co.jp)
	(2) Turkey. (Website: www.ntv.com.tr)
	(3) Newfoundland Television (Canada). (Website: www.ntv.ca)
NTVN	National Television Network (Tanzania)
NZBC	New Zealand Broadcasting Corporation (New Zealand)
ORF	Österreichischer Rundfunk Fernsehen (Austria). (Website: www.orf.at)
ORTB	Office de Radiodiffusion et de Télévision du Benin (Benin)
ORTS	Office de Radiodiffusion Télévision de Senegal (Senegal)
OTR	Old Time Radio (USA – radio)
OTV	Ostakino Television (Estonia-Russian language)
OZTV	Zaire Television (Zaire)
PBC	Pakistan Broadcasting Corporation (Pakistan)
PBS	(1) Philippine Broadcasting Service (Philippines)
	(2) Public Broadcasting System (USA). (Website: www.pbs.org)

PRBC	People's Revolution Broadcasting Corporation (Libya)
PTV	(1) Pakistan Television Corporation Limited (Pakistan). (Website: www.ptv.com.pk/index.htm)
	(2) Philippines Television (Philippines)
QTBS	Qatar Television and Broadcasting Service (Qatar)
QVC	Quality, Value, Choice (Website: UK: www.qvcuk.com, USA: www.qvc.com)
RAF	Radio Afghanistan (Afghanistan)
RAI	Radiotelevisione Italiana (Italy). (Website: www.rai.it/raiuno)
RBS	Rede Brasil Sul do Sul de Comunicacoes (Brazil). (W ebsite: www.rbstv.com.br)
RCA	Radio Corporation of America (USA). (Website: www.rca.com)
RCN-TV	Radio Cadena Nacional (Colombia). (W ebsite: www.rcn.com.co/rcntv/indice.htm)
RTCI	Rajawali Citra Televisi Indonesia (Indonesia). (W ebsite: www.rcti.co.id)
RDTC	RadioDiffusion-Television Centrafricaine (Central African Republic)
RFO	Réseau France d'Outre-mer (Malakoff, France)
RGT	Rede Globo de Televisao (Brazil). (Website: www.redeglobo.com.br)
RHK	Radio Hong Kong (Hong Kong)
RMC	Radio Monte Carlo (Monaco)
RNE	(1) Radio Nacional de Espana (Spain)
	(2) Radio Nepal (Nepal)
RNL	Radiodiffusion Nationale Lao (Laos)
RNW	Radio Nederlande Wereldomroep (Holland – radio)
RNZ	Radio New Zealand (New Zealand)
RPA	Televisao Popular de Angola (Angola)
RPC	Corporacion Panamena de Radiodifusion (Panama). (Website: www.rpctv.com)
RSR	Radio Suisse Romande (Switzerland – radio, Romanch language)
RTA	Radiodiffusion-Télévision Algerienne (Algeria)
RTB	(1) Radiodiffusion-Télévision Belge (Belgium)
	(2) Radio-Television Brunei (Brunei)
RTBF	Radio-Télévision Belge de la Communaut é culturelle Française (Belgium, French language). (W ebsite: www.rtbf.be)
RTD	Radiodiffusion Television de Djibouti (Djibouti)
RTE	Radio Telefis Éireann (Eire). (Website: www.rte.ie)
RTG	(1) Radio Televison Gabonaise (Gabon)

RTG	(2) Radio-Televison Guatemala (Guatemala)
cont.	(3) Radiodiffusion Televison Guineenne (Guinea)
RTH	Radio Thailand (Thailand)
RTHK	Radio-Television Hong Kong. (Website: www.rthk.org.hk)
RTI	Radiodiffusion Télévision Ivoirienne (Ivory Coast)
RTK	Radju ta' Kulhadd (Malta-radio)
RTL	Radio Télévision Luxembourg (Luxembourg)
RTM	(1) Radiodiffusion Télévision Marocaine (Morocco)
	(2) Radio Television Malaysia (Malaysia)
RTP	(1) Radiotelevisão Portuguesa (Portugal)
	(2) Radio Television Peruana (Peru)
RTR	(1) Radio e Televisiun Rumantscha (Switzerland, Romanch language)
	(2) Raioteleviziunea Romana (Romania)
RTS	Radio Television Singapore (Singapore)
RTSI	Radio e Televisione Svizzera di lingua Italiana (Switzerland, Italian language)
RTT	Radiodiffusion Télévision Tunisienne (Tunisia)
RTV	(1) Radiotelevision Limited (Hong Kong)
	(2) Reklaamitelevisioon (Estonia)
RTVA	Radiodiffusion-Television Algerienne (Algeria)
RTVE	Radio Television Espanola (Spain). (Website: www.rtve.es)
RTVS	Radio Televizija Solvenija (Slovenia). (Website: www.rtvs.si)
RTVSH	Radiotelevisione Shqiptar (Albania)
RUV	Rikisutvarpid (Iceland)
S4C	Sianel Pedwar Cymru (Wales). (Website: www.s4c.co.uk)
SABC	South African Broadcasting Corporation (South Africa). (Website: www.sabc.co.za)
SATV	Saudi Arabian Television (Saudi Arabia)
SBA	Singapore Broadcasting Authority (Singapore)
SBC	(1) Seychelles Broadcasting Corporation (Seychelles)
	(2) Swiss Broadcasting Corporation (= SRG, SSR, Switzerland). (Website: www.srg-ssr.ch)
SBS	(1) Scandinavian Broadcast Systems (based in London, UK)
	(2) Special Broadcasting Service (Australia). (Website: www.sbs.com.au)
SBT	Sistema Brasileiro de Televisão (Brazil). (Website: www.sbt.com.br)
SCTV	Surya Citra Television (Indonesia). (Website: www.sctv.co.id)
SDN	S4C Digital Networks (UK). (Website: www.sdn.co.uk)
SDR	Süddeutscher Rundfunk (Germany)

SES	Société Européene des Satellites (Luxembour g). (Website: www.ses-astra.com)
SF1	Schweizen Fernsehen 1 (Switzerland, German language)
SF2	Schweizen Fernsehen 2 (Switzerland, German language)
SIC	Sociedade Independente de Comunicacao (Portugal)
SLBC	Sri Lanka Broadcasting Corporation (Sri Lanka)
SLTV	Sierra Leone Television (Sierra Leone)
SMG	Scottish Media Group. (Website: www.smg.com)
SRC	Société Radio-Canada. (Website: www.src-mtl.com/src-tv/intro.htm)
SRG	Schweizerische Rundpruch Gesellschaft (= SBC, SSR, Switzerland)
SRI	Swedish Radio International (Sweden)
SRT	Sveriges Radio (Sweden)
STAR	Satellite Television Asian Region (Asia)
STER	Stichting Ether Reclame (Holland)
STMB	Sistem Televisyen Malaysia Berhad (Malaysia)
STV	(1) Saeta Television (Uruguay). (Website: www.multi.com.uy/canal10)
	(2) Satellite TV (Barbados)
	(3) Scottish Television (UK). (Website: www.scottishtv.co.uk)
	(4) Slovenska Televizia (Slovak Republic)
	(5) Sudan Television (Sudan)
	(6) Sverige Television. Subsidiary of Sveriges Radio (Sweden)
STVS	Suriname Televisie Stiching (Suriname)
STV12	Singapore Television Twelve. (Website: www.stv12.com.sg)
SVT	Sveriges Television (Sweden). (Website: www.svt.se)
TBC	(1) Tanzania Broadcast Commission (Tanzania)
	(2) Tonga Broadcasting Commission (Tonga)
TBN	Trinity Broadcast Network (USA). (Website: www.tbn.org)
TBS	(1) Tokyo Broadcasting System (Japan). (Website: www.tbs.co.jp)
	(2) Turner Broadcasting System, Inc. (USA). (Website: http://turner.com)
TCM	Turner Classic Movies (TV channel) (USA). (Website: http://tcm.turner.com)
TCS	Television Corporation of Singapore (Singapore). (Website: http://tcs.com.sg)
TDF	Télédiffusion de France (France)
TDN	Televicentro de Nicaragua, S.A. (Nicaragua). (Website: www.canal2.com.ni)
TF 1	Télévision Française 1 (France)

TFC	The Filipino Channel (Philippines)
THT	Tvoyo Novoye Televidiye (Russia, private commercial broadcaster based in Moscow)
TKP	Telewizynja Korporacja Partycypacyjna (Poland)
TMC	Tele Monte Carlo
TnaG	Telefis na Gaile (Eire)
TNB	Television Nationale du Burundi (Burundi)
TNDH	Televison Nationale D'Haiti (Haiti)
TNT	Turner Network Television (USA – cable TV). (Website: http://tnt.turner.com)
TPI	Televisi Pendidikan Indonesia (Indonesia)
TPS	Télévision Par Satellite (France)
TQS	Télévision Quatre Saisons (Canada)
TROS	Televisie Radio Omroep Stichting (Holland). (W ebsite: www.omroep.nl/tros)
TRR	Televisiun Rumantscha (Switzerland, Romansch language). (Website: www.srg-ssr.ch/TRR)
TRT	(1) Tamil Radio & TV (Sri Lanka)
	(2) Turkish Radio Television Corporation (Turkey)
TS	Tele-Sahel (Niger)
TSI	Televisione svizzera di Lingua Italiana (Switzerland, Italian language). (Website: www.srg-ssr.ch/TSI)
TSN	The Sports Network (USA). (W ebsite: www.sportsnetwork.com/home.asp)
TSS	Televidenie Sovietskoio-Socivsa (Russian Federation)
TSW	Television South-West (UK). (Hist.)
TTT	Trinidad and Tobago Television (Trinidad)
TTV	Taiwan Television Enterprise (Taiwan)
TVB	(Hong Kong)
TVE	TV Espanola (Spain). (Website: www.rtve.es)
TVI	(1) TV Independent (Portugal). (Website: www.tvi.pt)
	(2) Television Itapua (Paraguay)
TVN	(1) Televisora Nacional (Panama)
	(2) Television Nacional de Chile (Chile). (W ebsite: http://iusanet.cl/tvn/tvn.htm)
TVNC	Television Northern Canada. (Website: www.tvnc.ca)
TVNZ	Television New Zealand (New Zealand). (W ebsite: www.tvnz.co.nz)
TVO	Television Ontario (Canada)
TVP	Telewizja Polska (or Polska Radio Telewizja) (Poland)
TVRI	Televisi Republik Indonesia (Indonesia)
TVS	Televizija Sarajevo (Bosnia-Herzegovina)

TVSF	TV Schweizer Fernsehen DRS (Switzerland, German language)
TVSR	Télévision Suisse Romande (Switzerland, French language). (Website: www.tsr.srg-ssr.ch)
TVT	Television Togolaise (Togo)
TVV	Television Vietnam (Vietnam)
TVZ	Television Zanzibar (Tanzania)
TV2	(1) Television 2 (Denmark). (Website: www.dknet.dk/tv/tv2) (2) Television 2 (New Zealand). (W ebsite: www.tv2.co.nz)
TV3	Television 3 (New Zealand). (W ebsite: www.tv3.co.nz)
TV4	(1) Television 4 (New Zealand). (W ebsite: www.tv4.co.nz) (2) Television 4 (Sweden). (Website: www.tv4.se)
TWC	Time Warner Cable (USA). (W ebsite: www.timewarner.com/corp/about/cablesys/twcable/index.html)
UAETV	United Arab Emirates Televison (United Arab Emirates)
UBC	United Broadcasting Corporation (Thailand)
UCV	Universidad Catolica de Valpararaiso (Chile)
UPN	United Paramount Network (USA). (W ebsite: www.upn.com/hmupn.htm)
UR	Utbildningsradion (Sweden). (Website: www.ur.se)
USSB	United States Satellite Broadcasting (USA). (W ebsite: www.ussbtv.com)
UTB	Ukrajinska Telebacennja (Ukraine)
UTV	(1) Uganda Television (Uganda) (2) Ulster Television (UK). (Website: www.utvlive.com) (3) Uzbek Television (Uzbekistan)
VOA	Voice Of America (USA – radio)
VOK	Voice Of Kenya (Kenya – radio)
VON	Voice Of Nigeria (Nigeria – radio)
VOO	Veronica (Holland). (Website: www.veronica.nl)
VPRO	Vrijzinnig-Protestantse Radio Omroep (Holland). (W ebsite: www.vpro.nl)
WB	Warner Brothers Television Network (USA). (W ebsite: www.thewb.com)
WDR	Westdeutscher Rundfunk (Germany). (W ebsite: www.wdr.de)
XEW	Mexico. (Website: www.televisa.com/television/canal2/index.html)
YLE	Oy Yleisradio Ab (Finland). (Website: www.yle.fi)
YRTV	Yemen Radio & TV Corp. (Yemen)
YTTTV	Yorkshire Tyne Tees Television (UK). (Website: www.granadamedia.com)

Appendix Q: World television broadcasters

YTV   Yorkshire Television (UK). (Website: www.granadamedia.com)

ZDF   Zweites Deutsches Fernsehen (Germany). (Website: www.zdf.de)

ZTV   (1) Norway. (Website: www.ztv.no)
               (2) Zimbabwe Television (Zimbabwe)

Appendix R
Analogue television broadcast systems

Two sets of parameters are defined by the *International Telcommunications Union* in ITU-R Recommendation BT.470 (see **Appendix V**) for analogue television services:

1 Colour encoding systems, falling into three main categories (see **Table App.R.1**).
2 Scanning and modulation parameters given single letter designations (see **Table App.R.2**).

These are added together to describe a particular service, e.g. PAL-I for the UK and SECAM-L for France. See also **Appendix I**.

Table App.R.1 Colour television encoding system parameters

Parameter	PAL (B,C,G,H)	PAL (I)	PAL (M)	PAL (N)	NTSC (M)	SECAM (D,K,L)
Subcarrier (MHz)	4.433 618 75	4.433 618 75	3.575 611 49	3.582 056 25	3.579 545	$f_{Dr} = 4.406\,250$ $f_{Db} = 4.250\,000$
Subcarrier tolerance (Hz)	± 5	± 1	± 10	± 5	± 10	± 2000
Chrominance modulation	Suppressed carrier AM	Suppressed carrier AM	Suppressed carrier AM	Suppressed carrier AM	Suppressed carrier AM	FM
Burst amplitude (mV)	300	300	300	300	286 (40 IRE)	–
Burst duration (nominal)	10 cycles	10 cycles	9 cycles	9 cycles	9 cycles	–
Notes			Only used in Brazil	Only used in Argentina		

Table App.R.2 World analogue television systems

Parameter	System code												
	A	B	C	D	E	G	H	I	K	L	M (PAL)	M (NTSC)	N
Lines per picture	405	625	625	625	819	625	625	625	625	625	525	525	625
Field frequency (Hz)	50	50	50	50	50	50	50	50	50	50	60	59.94	50
Line frequency (Hz)	10 125	15 625	15 625	15 625	20 475	15 625	15 625	15 625	15 625	15 625	15 750	15 734	15 625
Video bandwidth (MHz)	3	5	5	6	10	5	5	5.5	6	6	4.2	4.2	4.2
Channel bandwidth (MHz)	5	7	7	8	14	8	8	8	8	8	6	6	6
Analogue sound carrier spacing from vision carrier (MHz)	-3.5	+5.5	+5.5	+6.5	+11.15	+5.5	+5.5	+6	+6.5	+6.5	+4.5	+4.5	+4.5
Vestigial sideband (MHz)	0.75	0.75	0.75	0.75	2.0	0.75	1.25	1.25	1.25	1.25	0.75	0.75	0.75
Vision modulation polarity	+ve	-ve	+ve	-ve	+ve	-ve	-ve	-ve	-ve	+ve	-ve	-ve	-ve
Analogue sound modulation	am	fm ±25 kHz	am	fm ±50 kHz	am	fm ±50 kHz	fm ±50 kHz	fm ±50 kHz	fm ±50 kHz	am	fm ±25 kHz	fm ±25 kHz	fm ±25 kHz
FM pre-emphasis (µs)	—	75	50	50	—	50	50	50	50	—	75	75	75

Appendix S
Digital television broadcast systems

The three main digital television broadcasting systems being adopted around the world are listed below and in the table. See also**Appendix I**.

DVB (Digital Video Broadcasting)

The system was developed in Europe and based on MPEG-2 standards (see **Appendix X**). It has three main parts with a variety of possible picture formats:

- **DVB-C:** DVB framing, channel coding and modulation scheme for *cable television* systems (ETSI standard EN 300 429).
- **DVB-S:** System for digital satellite television (ETSI standard EN 300 421).
- **DVB-T:** System for *digital terrestrial television* using *coded orthogonal frequency division multiplexing (COFDM)* multiple-carrier transmission (ETSI standard EN 300 744).

More details on www.dvb.org.

ATSC (Advanced Television Systems Committee)

The ATSC in the USA developed this ATV (Advanced Television) system also called the Digital Television Standard. It is based on the MPEG-2 video and systems standards (see **Appendix X**) and the AC-3 digital audio compression system. It offers a wide range of video resolutions and audio services for terrestrial and cable transmission and is defined in standards ATSC Standards A/52 and A/53, and SMPTE 274M, 293M, 294M, 295M and 296M (see **Appendix U**).
More details on www.atsc.org.

ISDB (Integrated Services Digital Broadcasting)

Proposed by NHK and adopted by Japan for digital terrestrial television services (as ISDB-T). It is based on MPEG-2 standards (see **Appendix X**) together with *Hi-Vision*, the analogue HDTV satellite television service broadcast in Japan. ISDB services require an *Integrated Services Television (ISTV)* to enable services including fax, enhanced teletext, telemusic, PCM sound broadcasting and programme guides to all be offered through a single television.

ISDB-T uses a band segmented transmission – *orthogonal frequency division multiplexing* (BST-OFDM) scheme where a TV channel is split into 432 kHz segments. Modulation and error correction can be set independently for each segment.

More details on www.nhk.or.jp/index-e.html and www.dibeg.org.

Notes on Table App.S.1

1 DVB allows luminance resolutions of 720, 544, 480 or 352 pixels per line. (Sometimes called D1, $\frac{3}{4}$D1, $\frac{2}{3}$D1 and $\frac{1}{2}$D1).

2 Not every option is possible in ATSC. There are 18 combinations which are allowed, each with two frame rates.

3 Some US operators use a proprietary digital satellite broadcasting system called DSS® *(Digital Satellite System)*. It is based on MPEG-2 video and audio coding, but uses 544 pixels × 480 lines, non-standard 130 byte transport stream packets with proprietary scrambling (MPEG specifies 188 byte packets) and incompatible error protection. Its channel bandwidth is 24 MHz.

4 ISDB uses MPEG AAC (Advanced Audio Coding) developed for MPEG-4, which is not backwards compatible with MPEG Layer II audio.

Table App.S.1 Digital television broadcasting systems

Parameter	DVB			ATSC		ISDB
	DVB-T	DVB-S	DVB-C	ATSC-T	ATSC-C	ISDB-T
Video coding SDTV service HDTV service	MPEG-2 MP@ML MP@HL	MPEG-2 MP@ML MP@HL	MPEG-2 MP@ML MP@HL	MPEG-2 MP@ML MP@HL	MPEG-2 MP@ML MP@HL	MPEG-2 MP@ML MP@HL
Audio coding	MPEG-2 Layer II	MPEG-2 Layer II	MPEG-2 Layer II	AC-3	AC-3	MPEG AAC (ISO/IEC13818–7)
System coding	MPEG-2	MPEG-2	MPEG-2	MPEG-2	MPEG-2	MPEG-2
Transmission	COFDM	PSK	QAM	8 VSB	16 VSB	BST-OFDM
Modulation	QAM	QPSK	16 QAM, 64 QAM	8-level symbols	16-level symbols	QPSK, 16 QAM, 64 QAM, DPSK
No. of carriers	1705 (2k) 6817 (8k)	1	1	1	1	1405 (Mode 1) 2809 (Mode 2) 5617 (Mode 3)
Mobile reception	Yes	–	–	No	–	Yes

	Yes (8k system)	—	—	No		Yes
Single Frequency Network (SFN)	Yes (8k system)	—	—	No		Yes
Channel bandwidth (MHz)	6, 7, 8	26 to 72	6, 7, 8	6	6	6, 7, 8
Data throughput (Mbit/s)	3.7–23.8 (6 MHz channel) 4.9–31.7 (8 MHz channel)	Up to 39	Up to 39	19.3	38.6	Up to 23 (6 MHz channel) Up to 31 (8 MHz channel)
Pixels/line:						
4:3 SDTV	720	720	720	640, 704	640, 704	720
16:9 SDTV	720	720	720	704	704	720
16:9 EDTV	960	960	960	–	–	–
16:9 HDTV	1920	1920	1920	1280, 1920	1280, 1920	1920
Lines/frame:						
SDTV	576	576	576	480	480	480 (i or P)
EDTV	576	576	576	–	–	–
HDTV	1080	1080	1080	720, 1080	720, 1080	1080 (i only)
Frame rate & scanning: (i = interlaced, P = Progressive)	25i (SDTV) 25P (HDTV) 50P (HDTV)	25i 25P 50P	25i 25P 50P	24P, 23.976P, 30P, 29.97P, 30i, 29.97i, 60i, 59.94i	24P, 23.976P, 30P, 29.97P, 30i, 29.97i, 60i, 59.94i	29.97i 29.97P 59.94P

Appendix T
Standards bodies

AES	Audio Engineering Society (USA). (Website: www.aes.org)
ANSI	American National Standards Institute (USA). (Website: www.ansi.org)
ASA	American Standards Association (USA)
ATSC	Advanced Television Systems Committee (USA). (Website: www.atsc.org)
BSI	British Standards Institution (UK). (Website: www.bsi.org)
CENELEC	Comité Européenne de Normalization Electrotechnique (European Committee for Electrotechnical Standardisation). (Website: www.cenelec.be)
CIE	Commission Internationale de l' Éclairage. (Website: www.cie.co.at/cie/home.html)
DIN	Deutsches Institut für Normung (Germany). (Website: www.din.de)
DVB	Digital Video Broadcasting. (Website: www.dvb.org)
EBU	European Broadcasting Union. (Website: www.ebu.com)
EIA	Electronic Industries Association (USA). (Website: www.eia.org)
EIAJ	Electrical Industries Association of Japan (Japan). (Website: www.eiaj.or.jp/english/index.htm)
ETSI	European Telecommunications Standards Institute. (Website: www.etsi.org/eds)
IEC	International Electrotechnical Commission (works in the fields of electrical and electronic engineering). (Website: www.iec.ch)
IEE	Institute of Electrical Engineers (UK). (Website: www.iee.org.uk)
IEEE	Institute of Electrical and Electronic Engineers (USA). (Website: www.ieee.org)
ISO	International Organization for Standardization (all technical fields except electrical and electronic engineering). (Website: www.iso.ch)
ITU	International Telecommunications Union (sets standards in the field of telecommunications). (Website: www.itu.int)

JEDEC   Joint Electronic Devices Engineering Council.
(Establishes standards for the electronics industry .)

NAB   National Association of Broadcasters (USA). (W ebsite:
www.nab.org)

NBS   National Bureau of Standards (USA)

RIAA   Recording Industries Association of America (USA).
(Website: www.riaa.com)

SMPTE   Society of Motion Picture and Television Engineers
(USA). (Website: www.smpte.org)

Appendix U
SMPTE standards

Those standards referred to in the dictionary are listed here. The complete listing is available on the Internet at www .smpte.org. The SMPTE also publish Recommended Practice documents (RPs) and Engineering Guidelines (EGs), also listed on the website. New American (ANSI/SMPTE) standards are regularly published in the *Journal of the Society of Motion Picture and Television Engineers.*

Film

SMPTE 7–1999: Motion-Picture Film (16-mm) – Camera Aperture Image and Usage

SMPTE 40–1997: Motion-Picture Film (35-mm) – Release Prints – Photographic Audio Records

SMPTE 41–1999: Motion-Picture Film (16-mm) – Prints – Photographic Audio Records

SMPTE 48–1995: Motion-Picture Film (16-mm) – Picture and Sound Contact Printing – Printed Areas

SMPTE 55–1992: Motion-Picture Film – 35- and 16-mm Audio Release Prints – Leaders and Cue Marks

SMPTE 59–1998: Motion-Picture Film (35-mm) – Camera Aperture Images and Usage

SMPTE 75M–1994: Motion-Picture Film – Raw Stock – Designation of A and B Windings

SMPTE 83–1996: Motion-Picture Film (16-mm) – Edge Numbers – Location and Spacing

SMPTE 86–1996: Motion-Picture Film – Magnetic Audio Records – Two, Three, Four and Six Records on 35-mm and One Record on 17.5-mm Magnetic Film

SMPTE 152–1994: Motion-Picture Film (70-mm) – Projectable Image Area

SMPTE 185–1993: Motion-Picture Film (70-mm) – Six Magnetic Records on Release Prints – Position, Dimensions, Reproducing Speed and Identity

SMPTE 195–1993: Motion-Picture Film (35-mm) – Motion-Picture Prints – Projectable Image Area

SMPTE 196M–1995: Motion-Picture Film – Indoor Theater and Review Room Projection – Screen Luminance and Viewing Conditions

SMPTE 201M–1996: Motion-Picture Film (16-mm) – Type W Camera Aperture Image

SMPTE 203–1998: Motion-Picture Film (35-mm) – Prints – Two-Track Photographic Audio Records

SMPTE 215–1995: Motion-Picture Film (65-mm) – Camera Aperture Image

SMPTE 221–1998: Motion-Picture Film (70-mm) – Six-Track Audio Release Prints – Magnetic Striping

SMPTE 233–1998: Motion-Picture Film (16-mm) – Projectable Image Area and Projector Usage

Television

SMPTE 12M–1995: Television, Audio and Film – Time and Control Code

SMPTE 96–1992: Television – 35- and 16-mm Motion-Picture Film and 2×2-in Slides – Scanned Area and Photographic Image Area for 4 :3 Aspect Ratio

SMPTE 170M–1999: Television – Composite Analog Video Signal – NTSC for Studio Applications

SMPTE 240M–1995: Television – Signal Parameters – 1125-Line High-Definition Production Systems

SMPTE 259M–1997: Television – 10-Bit 4 :2:2 Component and 4fsc Composite Digital Signals – Serial Digital Interface

SMPTE 260M–1999: Television – 1125/60 High-Definition Production System – Digital Representation and Bit-Parallel Interface

SMPTE 266M–1994: Television – 4:2:2 Digital Component Systems – Digital Vertical Interval Time Code

SMPTE 274M–1998: Television – 1920 × 1080 Scanning and Analog and Parallel Digital Interfaces for Multiple Picture Rates

SMPTE 277M–1996: Television Digital Recording – 19-mm Type D-6 – Helical Data, Longitudinal Index, Cue and Control Records

SMPTE 279M–1996: Digital Video Recording – 1/2-in Type D-5 Component Format – 525/60 and 625/50

SMPTE 292M–1998: Television – Bit-Serial Digital Interface for High-Definition Television Systems

SMPTE 293M–1996: Television – 720 × 483 Active Line at 59.94-Hz Progressive Scan Production – Digital Representation

SMPTE 294M–1997: Television – 720 × 483 Active Line at 59.94-Hz Progressive Scan Production – Bit-Serial Interfaces

SMPTE 295M–1997: Television – 1920 × 1080 50-Hz – Scanning and Interface

SMPTE 296M–1997: Television – 1280 × 720 Scanning, Analog and Digital Representation and Analog Interface

SMPTE 297M–1997: Television – Serial Digital Fiber Transmission System for ANSI/SMPTE 259M Signals

SMPTE 305M–1998: Television – Serial Data Transport Interface

SMPTE 306M–1998: Television Digital Recording – 6.35-mm Type D-7 Component Format – Video Compression at 25 Mb/s – 525/60 and 625/50

SMPTE 314M–1999: Television – Data Structure for DV -Based Audio, Data and Compressed Video – 25 and 50 Mb/s

SMPTE 316M–1999: Television Digital Recording – 12.65-mm Type D-9 Component Format – Video Compression – 525/60 and 625/50

SMPTE 318M–1999: Television and Audio – Reference Signals for the Synchronization of 59.94- or 50-Hz Related Video and Audio Systems in Analog and Digital Areas

Appendix V
ITU standards

ITU-R Recommendations – Series Subjects

ITU-R Recommendations are a set of technical and operating standards for radiocommunications, previously known as CCIR Recommendations. They are the result of studies undertaken by Radiocommunication Study Groups and are divided into Series Subjects according to the list below . A full list of ITU-R Recommendations can be found on the ITU website at www.itu.int/publications/itu-r/itur.htm.

BO	Broadcasting-satellite service (sound and television)
BR	Sound and television recording
BS	Broadcasting service (sound)
BT	Broadcasting service (television)
F	Fixed service
IS	Inter-service sharing and compatibility
M	Mobile, radiodetermination, amateur and related satellite services
P	Radiowave propagation
RA	Radioastronomy
S	Fixed-satellite service
SA	Space applications and meteorology
SF	Frequency sharing between the fixed-satellite service and the fixed service
SM	Spectrum management techniques
SNG	Satellite news gathering
TF	Time signals and frequency standards emissions
V	Vocabulary and related subjects

BT series examples

BT.470–4 (10/95)	Television systems
BT.500–7 (10/95)	Methodology for the subjective assessment of the quality of television pictures
BT.601–5 (10/95)	Studio encoding parameters of digital television for standard 4:3 and wide-screen 16:9 aspect ratios
BT.656–3 (10/95)	Interfaces for digital component video signals in 525-line and 625-line television systems operating at the 4:2:2 level of Recommendation ITU-R BT .601 (Part A)

ITU-T Recommendations – List of Series

There are 15 ITU-T Study Groups who together publish recommendations in the following series:

Series A	Organization of the work of the ITU-T
Series B	Means of expression
Series C	General telecommunication statistics
Series D	General tariff principles
Series E	Telephone network and ISDN
Series F	Non-telephone telecommunication services
Series G	Transmission systems and media
Series H	Transmission of non-telephone signals
Series I	Integrated services digital network
Series J	Transmission of sound-programme and television signals
Series K	Protection against interference
Series L	Construction, installation and protection of cables and other elements of outside plant
Series M	Maintenance: international transmission systems, telephone circuits, telegraphy, facsimile and leased circuits
Series N	Maintenance: international sound-programme and television transmission circuits
Series O	Specifications of measuring equipment
Series P	Telephone transmission quality
Series Q	Switching and signalling
Series R	Telegraph transmission
Series S	Telegraph services terminal equipment
Series T	Terminal equipments and protocols for telematic services
Series U	Telegraph switching
Series V	Data communication over the telephone network
Series X	Data networks and open system communication
Series Z	Programming languages

V series examples

ITU-T V.34	describes data communications via a modem at up to 28 800 bit/s
ITU-T V.34+	describes data communications via a modem at up to 33 600 bit/s
ITU-T V.90	describes data communications via a modem at up to 56 kbit/s

Appendix W
Useful Internet websites

References to websites are made throughout the dictionary. Here are some additional websites which are useful sources of information.

www.aes.org/resources/www-links   AES links to audio industry related sites
www.bksts.com/links   BKSTS links to film, TV and audio related sites
www.oldradio.com   Broadcast Archive
www.broadcast.net   BNet, a US broadcast industry site
www.DigitalTelevision.com/index   A US source of digitalTV information
english-server.hss.cmu.edu/Film&TV   Film & Television Archive
www.neog.com/mbc   Museum of Broadcast Communications, the history of radio and TV
www.newsworld.co.uk   The News World website is a meeting place for the news industry.
www.mpeg.org/mpeg/news   A US source of MPEG related information
www.nmsi.ac.uk/nmpft   The National Museum of Photography, Film & Television (Bradford, UK)
tvnet.com   All US TV stations
www.UltimateTV.com   World-wide TV related site
www.wssn.net/WSSN   World Standards Services Network, a source of standards information
www.yahoo.com/Entertainment/Movies_and_Films   Yahoo's index of movie related information

Film studios

www.movies.com   Buena Vista/Disney/Hollywood Pictures/Touchstone
www.spe.sony.com/movies/index   Sony, Columbia, Tri-Star
www.flf.com:80/index   Fine Line Films
www.reellife.com/PFE   Gramercy Pictures/PolyGram Filmed Entertainment
www.universalpictures.com   Universal Pictures
www.mgmua.com   MGM/UA
www.miramax.com   Miramax Films
www.newline.com   New Line Cinema
www.paramount.com   Paramount
www.foxmovies.com   Twentieth Century Fox
www.disney.com/DisneyPictures/?GL=H   Walt Disney Pictures
www.movies.warnerbros.com   Warner Bros

Appendix X
ISO/IEC MPEG standards

MPEG-1

ISO/IEC 11172   Information technology – Coding of moving pictures and associated audio for digital storage media at up to about 1.5 Mbit/s

ISO/IEC 11172-1:1993   Part 1: Systems
ISO/IEC 11172-2:1993   Part 2: Video
ISO/IEC 11172-3:1993   Part 3: Audio
ISO/IEC 11172-4:1995   Part 4: Compliance testing
ISO/IEC TR 11172-5   Part 5: Software simulation

MPEG-2

ISO/IEC 13818   Information technology – Generic coding of moving pictures and associated audio information

ISO/IEC 13818-1:1996   Part 1: Systems
ISO/IEC 13818-2:1996   Part 2: Video
ISO/IEC 13818-3:1998   Part 3: Audio
ISO/IEC 13818-4:1998   Part 4: Conformance testing
ISO/IEC TR 13818-5:1997   Part 5: Software simulation
ISO/IEC 13818-6:1998   Part 6: Extensions for DSM-CC is a full software implementation
ISO/IEC 13818-7:1997   Part 7: Advanced Audio Coding (AAC)
ISO/IEC 13818-9:1996   Part 9: Extension for real time interface for systems decoders
ISO/IEC 13818-10:1999   Part 10: Conformance extensions for Digital Storage Media Command and Control (DSM-CC)

MPEG-4

ISO/IEC 14496   Information technology – Coding of audio-visual objects

ISO/IEC 14496-1:1999   Part 1: Systems
ISO/IEC 14496-2:1999   Part 2: Visual
ISO/IEC 14496-3:1999   Part 3: Audio
ISO/IEC 14496-5:2000   Part 5: Reference software
ISO/IEC 14496-6:1999   Part 6: Delivery Multimedia Integration Framework (DMIF)

Appendix Y
DVD formats

DVD refers to a 12 cm digital disc which can have a variety of options for capacity and use, described below . It achieves a much higher storage capacity than *CD* by using a shorter wavelength *laser* beam, smaller pits and narrower track *pitch (1)*. Single-sided and double-sided versions are possible, with one or two layers of information on each side.

DVD-ROM

The basic pre-recorded DVD which can support DVD-V ideo and DVD-Audio formatted data. Its storage options are shown in **Table App.Y.1**.

Table App.Y.1   DVD capacities

Name	Structure	Capacity
DVD-5	Single-sided, single layer	4.7 gigabytes
DVD-9	Single-sided, dual layer	8.5 gigabytes
DVD-10	Double-sided, single layer	9.4 gigabytes
DVD-14	Double-sided, one single, one dual layer	13.2 gigabytes
DVD-18	Double-sided, double layer	17.1 gigabytes

DVD-Video

A pre-recorded DVD format which can be either single-sided, single layer (giving 120 minutes at 24 fps) or single-sided, dual layer. In the latter case, reverse spiral dual layer (RSDL) can give continuous replay of long movies.

- **Video:** DVD-Video supports 16:9 *aspect ratio* images at full resolution on widescreen television displays. For 4 :3 screens a DVD player can create a 16 :9 *letterboxed* image within the 4 :3 frame. DVD-V ideo movies use *variable bit rate (VBR)* to optimize quality and playing time. For example, a typical 24 fps movie has an average bit rate of 3.5 Mbit/s, but for action scenes with a great deal of movement the peak bit rate might be 8 or 9 Mbit/s.

Appendix Y: DVD formats

- **Audio:** DVD-Video supports *PCM, MPEG* and *Dolby Digital (AC–3)* audio with mono, stereo and *Dolby Surround* to 5.1 channels. At least one of these formats must be used, but DVD-V ideo discs can have others, including *Digital Theater Systems (DTS)* and *Sony Dynamic Digital Sound (SDDS)* . Up to eight separate audio streams can be used, allowing multiple language, audio description, director 's commentary, etc.
- **Region coding:** DVD-Video discs can be *region coded* to play only in one region or set of regions. They can also be 'code-free'. The region is defined in the player , so a *region coded* disc will only play on a player which is allowed by the coding. The region numbers are given in **Table App.Y.2**.

Table App.Y.2 DVD-Video region numbers

Number	Region
1	Canada, USA, USA territories
2	Japan, Europe, South Africa, Middle East (including Egypt)
3	Southeast Asia, East Asia (including Hong Kong)
4	Australia, New Zealand, Pacific Islands, Central America, South America, Caribbean
5	Former Soviet Union, Indian Subcontinent, Africa, North Korea, Mongolia
6	China
7	Reserved
8	Special international venues (airplanes, cruise ships, etc.)

DVD-RAM

A type of DVD which is erasable and can be rewritten. Single-sided discs have 2.6 or 4.7 gigabyte capacity , double-sided discs can store 5.2 or 9.4 gigabytes.

DVD-Audio

A pre-recorded DVD format to carry very high quality audio, sampled at up to 96 kHz. DVD-Audio is copyright protected by an embedded signalling or digital watermark feature.

DVD-AudioV

A type of DVD-Audio containing some additional video data in the DVD-Video standard.

Hybrid DVD

The term Hybrid DVD can have many meanings:

1 A DVD which plays differently on a DVD player and a DVD-ROM PC. It contains an executable program which takes advantage of PC features (e.g. enhanced graphics capabilities or Internet connectivity) which are not found on DVD players. More accurately called Enhanced DVD.
2 A DVD-ROM disc which runs on   Windows and Mac OS computers. More accurately called a Cross-platform DVD.
3 A DVD-ROM or DVD-Video disc which also contains Web content for connecting to the Internet. More accurately called a   WebDVD or Web-connected DVD.
4 A disc containing both DVD-V ideo and DVD-Audio content. More accurately called a Universal DVD.
5 A disc with two layers, one containing pressed DVD-ROM data, the other containing rewritable (DVD-RAM, etc.) media for recording and re-recording. More accurately called a Mixed-Media DVD or Rewritable Sandwich disc.
6 A disc with two layers on one side and one layer on the other   . More accurately called a DVD-14.
7 A disc with an embedded memory chip for storing custom usage data and access codes. More accurately called a Chipped DVD.

DVD+RW

A 3 gigabyte erasable and rewritable format proposed by Sony, Philips and Hewlett-Packard as an alternative to DVD-RAM.

DVD-R

A type of DVD which allows one time recording of data. DVD recordable discs can store 3.95 gigabytes on a single-sided disc and 7.9 gigabytes on a double-sided disc.

SACD (Super Audio CD)

A type of dual-format DVD developed by Sony and Philips with two layers, one that plays in existing CD players, plus a high-density layer for DVD-Audio players.

Appendix Z
Colour bar test signals and displays

Figure App.Z.1 shows a PAL vectorscope display of 100% P AL colour bars.

    Figure App.Z.2 shows a component vectorscope display of 100% colour bars. Note the display is circular rather than an ellipse and there is no burst vector.

    Figures App.Z.3, 4 and 5 show the line-rate waveforms for the three types of analogue composite P AL colour bars.

1 100% bars are only used in a broadcast production environment.
2 95% bars were used by the BBC (e.g. at the top of the   Test Card).
3 75% bars (or EBU bars) are used for circuit testing and identification between broadcasters.

Figure App.Z.6 shows a *parade display* of 625-line component 100% colour bars. A 350 mV offset is often used to align the peaks of the colour difference signals with black level and white level to make monitoring levels easier.

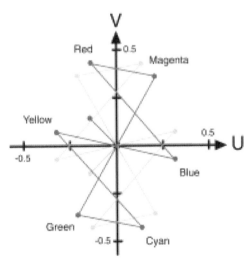

Figure App.Z.1   PAL vector display of 100% colour bars

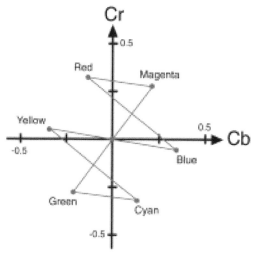

Figure App.Z.2 Component vector display of 100% colour bars

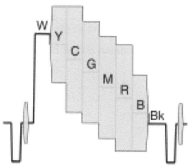

Figure App.Z.3 100% PAL colour bars

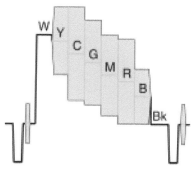

Figure App.Z.4 95% PAL colour bars

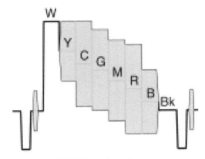

Figure App.Z.5   75% EBU PAL colour bars

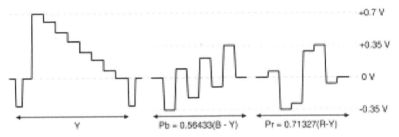

Figure App.Z.6   100% component colour bars